'The book is a revelation: The authors have done an impressive job in bringing together all the main areas of actual contemporary human–animal interaction research.'

Prof. Robert Mitchell, *Department of Psychology, Eastern Kentucky University, USA*

'Each chapter provides a really nice introductory text for key topics taught across a range of disciplines in HAI.'

Prof. Carri Westgarth, *Institute of Infection, Veterinary and Ecological Sciences, University of Liverpool, UK*

Introduction to Human–Animal Interaction

Introduction to Human–Animal Interaction focuses on the human dimension of interacting with other animals. This book introduces recent developments, theories and debates in the relatively new research area of human–animal interaction (HAI) and focuses on the social and life sciences aspect of these interactions. Experts from different academic disciplines provide an overview for students and professionals interested in how humans and other animals interact, and what advantages and disadvantages emerge for both parties in this relationship.

The book starts with the theories and mechanisms supporting our interactions with animals, such as human–animal communication, and it then covers the implications of HAI in terms of ethics and welfare. After discussing cultural differences and forensic aspects in HAI (e.g., wildlife crime and animal abuse), the book examines evidence in the area of animal-assisted intervention. The final chapters give an overview of current research in specific human–animal interaction systems: human–pet, human–livestock and human–wildlife interaction. The book offers a scientific, evidence-based perspective on human–animal interaction, providing pedagogical tools to make a systematic, critical and constructive evaluation of research in HAI possible. It offers a range of in-text pedagogical features like a subject index, chapter MCQs, open questions, further reading and additional digital resources including videos which are accessible via QR codes or through the associated website.

This textbook provides the fundamental tools for achieving a comprehensive, current and critical overview of the HAI field and is an integral text for undergraduate and postgraduate students undertaking modules in human–animal interaction, in social sciences such as anthropology, cultural studies, criminology, ethics and laws, or in life sciences such as animal behaviour, conservation and welfare, biology, neuroscience, physiology, psychology, public health and those studying veterinary sciences.

Laëtitia Maréchal is Associate Professor in Psychology at the University of Lincoln (UK), with a multidisciplinary background from the social and life sciences, including psychology, ethology, endocrinology, socio-ecology, primatology and social anthropology. Her research applies a One Health/One Welfare approach to better understand interspecific interactions. She is also a member of the International Union for the Conservation of Nature (IUCN) Primate Specialist Group Section for Human–Primate Interactions.

Emile van der Zee is Associate Professor in Psychology at the University of Lincoln (UK). He has published numerous books on language and published several articles and book chapters related to animal communication. He designed a popular course titled Psychology of Human–Animal Interaction, which inspired this book.

Introduction to Human–Animal Interaction

Insights from Social and Life Sciences

Edited by Laëtitia Maréchal and Emile van der Zee

LONDON AND NEW YORK

Designed cover image: © Loïc Maréchal

First published 2024
by Routledge
4 Park Square, Milton Park, Abingdon, Oxon OX14 4RN

and by Routledge
605 Third Avenue, New York, NY 10158

Routledge is an imprint of the Taylor & Francis Group, an informa business

British Library Cataloguing-in-Publication Data
A catalogue record for this book is available from the British Library

ISBN: 9781032118376 (hbk)
ISBN: 9781032118369 (pbk)
ISBN: 9781003221753 (ebk)

DOI: 10.4324/9781003221753

Typeset in Times New Roman
by Newgen Publishing UK

Contents

Editor and Author Biographies

Editors

Dr. Laëtitia Maréchal is Associate Professor in Psychology at the University of Lincoln (UK), with a multidisciplinary background from Social and Life Sciences, including Psychology, Ethology, Endocrinology, Socio-Ecology, Primatology and Social Anthropology. Her research applies the One Health/One Welfare approach to better understand interspecific interactions. She is also a member of the International Union for the Conservation of Nature (IUCN) Primate Specialist Group Section for Human–Primate Interactions.

Dr. Emile van der Zee is Associate Professor in Psychology at the University of Lincoln (UK). He has published numerous books on language and published several articles and book chapters related to animal communication. He designed a popular course titled 'Psychology of Human–Animal Interaction', which inspired this book.

Authors

Naeem Abbas is an experienced community development and public policy expert at the Brooke Hospital for Animals (Pakistan), currently working as Advocacy Manager with Brooke Pakistan. In his current role, he is working with government and civil society organisations to create enabling environment for animal welfare. His specialties are: policy reform, community engagement, research (qualitative and quantitative, participatory and action research), PRA/participatory learning for action methodologies, logical framework approach, project cycle management and strategy creation.

Dr. Ana Maria Barcelos is Postdoctoral Researcher and Guest Lecturer in human–animal interaction and human mental health at the University of Lincoln (UK), with a strong research background in the effects of pet ownership on owners' well-being. In addition, she is a Veterinarian and the President of the Brazilian Association of Veterinary Behaviourists. Thus, she also investigates animal behaviour and the impact of human–animal interactions on animal welfare.

Dr. Ross M. Bartels is Senior Lecturer in Forensic Psychology and leads the Forensic and Crime Research Group at the University of Lincoln (UK). His research focuses on understanding atypical sexual interests, the psychology of sexual fantasising and offence-supportive cognitions within forensic and community populations. Ross has published numerous papers and book chapters, has co-edited two books on sexual deviance, and regularly presents at forensic conferences.

Dr. Victoria Brelsford is Associate Lecturer in Psychology with an interdisciplinary background in early childhood development and human–animal interaction (HAI) research at the University of Lincoln (UK), including the effect of companion animals on cognition, physiological processing, and humans' understanding of canine stress signalling behaviour. She is passionate about promoting the welfare of all animals and teaching the critical importance of this to both children and adults.

Dr. Giovanna Capponi is Postdoctoral Researcher at the University of Roehampton (UK). She is trained as a social anthropologist with a particular interest in environmental anthropology, human–animal studies and more-than-human ontologies. She conducted research on animal sacrifice in the Afro-Brazilian Candomblé religion, on wild boar management and human–wildlife conflicts in Italy, and she is currently working on feral cat colonies in urban settings.

Justine Cole graduated with a Bachelor of Science in Applied Animal Biology Honours Program from the University of British Columbia (Canada). Justine's previous research has focused on zoo animals, wildlife and on highlighting the largely overlooked role that human–animal interactions play in both animal welfare and human well-being.

Prof. Grahame J. Coleman is Professor in Psychology at the University of Melbourne (Australia), conducting research on public attitudes to farm and companion animal welfare and research into the effects of human–animal interactions on farm animal welfare. He has also focused on the development of cognitive-behavioural interventions to change human attitudes and behaviour leading to improved production and welfare outcomes for farm animals.

Dr. Mirena Dimolareva is Lecturer in Educational Psychology with a background in child development and animal–assisted intervention research at the University of Lincoln (UK). She is interested in how we can adapt animal-assisted interventions for different groups of people within our society, particularly those from disadvantaged backgrounds or with additional needs.

Dr. Georgina Gous is Senior Lecturer in Forensic Psychology at the University of Lincoln (UK). She specialises in research at the pre-conviction stage of the criminal justice process, with such work including the use of voice disguise, leading questions in court, witness preparation techniques, the safeguarding of vulnerable witnesses, jury decision-making, and the reducing of juror biases. Georgina has published numerous research articles and contributed to several book chapters.

Prof. Kun Guo is Professor in Cognitive Neuroscience at the University of Lincoln (UK). He combines behavioural, physiological, neuroimaging and computational approaches to study social attention and affective vision (e.g., face and body perception, emotion) in both humans and non-human animals. Since 2009, his group has pioneered a range of neuroscientific methods (psychophysics, eye-tracking) to study how humans and dogs read each other's emotions and human–dog interaction.

Dr. Roxanne Hawkins is Lecturer in Applied Psychology at the University of Edinburgh (UK). She is a multidisciplinary researcher with expertise in human–animal interactions, with a focus on mental health implications of attachment to pets, animals within adversity, risk and resilience, animal welfare and prevention and intervention strategies for animal cruelty prevention.

She works with external organisations such as animal welfare charities and participates in a range of public engagement activities.

Prof. Paul H. Hemsworth is Professor at the University of Melbourne (Australia). His two main research programmes have focused on the effects of (1) the human–animal relationship on the welfare of farm, zoo and companion animals and (2) housing and husbandry practices on the welfare of farm animals. He has published more than 250 refereed journal articles, 35 chapters in internationally distributed books and one book on animal behaviour and welfare.

Prof. Catherine M. Hill is Professor in Anthropology at the Oxford Brookes University (UK), with a multidisciplinary background in Zoology, Psychology and Anthropology. In her research she adopts a biosocial approach to explore conflicts around wildlife and human–wildlife coexistence. She is a member of the International Union for the Conservation of Nature (IUCN) SSC Human–Wildlife Conflict & Coexistence Specialist Group and the IUCN Primate Specialist Group Section for Human–Primate Interactions.

Laura Kavata Kimwele is an experienced technical specialist in the field of community development at the Brooke Hospital for Animals (East Africa). She has worked with local and international organisations on livelihood, resilience, food security, climate change and animal welfare in the past 16 years, with valuable experience on project planning and management. She holds a Bachelor of Arts in Community Development, Daystar University, and is pursuing a Master of Project Planning and Management, University of Nairobi (Kenya).

Dr. Meagan King is Assistant Professor in Animal Welfare, with a background in Environmental Biology and Animal Welfare and Behaviour at the University of Manitoba (Canada). She has mostly studied dairy cow health and welfare in robotic milking herds. Most recently, she surveyed dairy farmer mental health and how it is linked with animal health and farm management. She has also worked with piglets, wildlife, fish and zooplankton.

Dr. Tracie McKinney is Senior Lecturer in Biological Anthropology at the University of South Wales (UK), with a focus on primate ecology, behaviour and conservation. Her research uses an ethnoprimatology framework to explore ways humans and non-human primates interact, in situations such as tourism, crop feeding and forest fragmentation. She is a member of the IUCN Primate Specialist Group Section for Human–Primate Interactions.

Prof. Kerstin Meints is Professor in Developmental Psychology, and Suffrage Women in Science Award holder at the University of Lincoln (UK). Her research spans children's development, with a focus on cognition and language as well as human–animal interaction, especially dog bite prevention and animal-assisted interventions. As director of the Lincoln Infant and Child Development Lab, head of Lincoln Education Assistance with Dogs (lead.blogs.lincoln.ac.uk), head of the Development and Behaviour Research group and Autism Research and Innovation Centre member, she encourages interdisciplinary research.

Prof. Ben Mepham is Honorary Professor at the University of Nottingham (UK), he established the Centre for Applied Bioethics, a teaching and research institute in 1993. He was director of the Food Ethics Council (1998–2005), appointed to the UK Government Biotechnology Commission (2000–2005), is co-founder of the European Society for Agricultural and Food

Ethics (appointed Life member on retirement), and is a member of several related European Union and UK committees. His *Bioethics: An Introduction for the Biosciences* (2005 and 2nd edition 2008) and other books (e.g., *Food Ethics*) have won significant acclaim.

Prof. Angus Nurse is Professor of Law and Environmental Justice at Anglia Ruskin University (UK), with research interests in green criminology, criminality, critical criminal justice, animal rights and human rights law. He is a member of the Wild Animal Welfare Committee (WAWC) and co-chair of the Wildlife Working Group for the UK Centre for Animal Law. His books include *Policing Wildlife* (2015) and *Animal Harm* (2013).

Dr. Ricardo R. Ontillera-Sánchez is a Postdoctoral Researcher at the University of Roehampton (UK). He is trained as a biologist and social anthropologist with a particular interest in human–animal studies. He conducted research on transhumance in mainland Spain, and he is currently the principal investigator of an ESRC-funded project (ES/W006472/1) which studies the cultural constructions of fighting birds and fighting bulls in the Canary Islands and Andalusia (Spain).

Daniela Pörtl, MD, is a Psychiatrist, Psychotherapist and Neurologist working as a Senior Consultant in the Department of Psychiatry, SRH Hospital Naumburg, teaching hospital of Jena and as well in my psychotherapeutic practice at the Leipzig Universities, Leipzig, Germany, with a multidisciplinary background from Neurobiology, Epigenetics Endocrinology and Social Anthropology. Her scientific work is about self-domestication processes, dog domestication, language evolution and epigenetic modulations in the brain due to prosocial behaviour and stress.

Dr. Melanie Ramasawmy is Postdoctoral Research Fellow at the University College London (UK) with a multidisciplinary background including Social Anthropology and Life Sciences, and experience in health and social care policy. She carried out ethnographic research on gender and agriculture in rural Ethiopia as part of her doctoral and postdoctoral research. Her current research uses mixed methods to understand inequalities in digital health.

Dr. Rie Usui 笛吹 理絵 is Assistant Professor at Ritsumeikan Asia Pacific University in Japan with a multidisciplinary background in ecology, anthropology, primatology and geography. Her main research interests lie in the overlapping area of tourism and human–animal studies. Her current research project focuses on the cases of liminal animals, such as feral rabbits and cats, that have become a tourism attraction.

Dr. Shelly Volsche is Clinical Assistant Professor in Anthropology at University of Wisconsin – River Falls (USA), with a biocultural and transdisciplinary background including Psychology, Animal Behaviour, Anthrozoology and Evolutionary Cognition. Her research uses the One Health/One Welfare lens to investigate behavioural health within and between species. She is a member of the International Society for Anthrozoology and the American Anthropological Association, as well as Human Behaviour and Evolution Society.

Dr. Jennifer Wathan is Global Animal Welfare Adviser at Brooke, Action for Working Horses and Donkeys (UK), an international animal welfare charity operating in Africa, Asia, Latin

America and the Middle East. Experienced in research, psychology and human behaviour change, she utilises a One Health/One Welfare approach that recognises the complex interaction between working animals, the communities who depend upon them, and the environment within which people and animals coexist.

Dr. Eva Zoubek is a multidisciplinary researcher who studied the small-scale chicken keeping culture in Britain and the human–animal relationships within this context as part of her doctoral studies from the University of Roehampton (UK). Previously, she studied environmental studies in Vienna and environment–human relationships at the University College London Department of Geography.

Acknowledgements

The creation of this textbook would not have been possible without the support of the publisher and editors, as well as the contributions of the authors. We also would like to thank Loïc Maréchal for the cover illustration, as well as the learning (istockphotos) and the QRcode drawings. We thank numerous reviewers whose comments have greatly improved this book.

1 An Introduction to Human–Animal Interaction, Book Content and Considerations

Laëtitia Maréchal and Emile van der Zee

1.1 Introduction

Our relationship with other animals is complex, ambiguous and ubiquitous. For instance, we share our environment with wildlife, and we keep animals for companionship, entertainment, research, sports and food, while using their products for manufacturing shoes, crayons, hormones, biofuel and beauty products (DeMello, 2012). Our ambiguous relationship with animals is expressed, for example, in keeping rabbits as companion animals, using them as food for ourselves and other animals, using them in manufacturing (wool, glue and felt), as well as retaining them as models for pharmaceutical and medical research. Our relationship with animals may stretch from an emotional bond that goes beyond a bond with members of our own species (e.g. feeling sexually attracted to animals – zoophilia) to using animals as instruments for any possible purpose we can think of (e.g. shooting them for recreational purposes, or using them to detect human earthquake victims and to explore space). While on the one hand we invest emotionally and financially in animals, on the other hand, we exploit animals for any possible resource their presence brings us. How can we explain this complex, ambiguous and ubiquitous relationship with other animals? What are the consequences of our relationship with other animals for ourselves, for those other animals, and even for the biota in general? For example, our interventions in animal species distribution, such as reintroducing wolves into Yellowstone National Park had an impact on the surrounding flora, fauna and even the climate (Smith & Peterson, 2021). It is these kinds of questions that the research area of human–animal interaction (HAI) addresses, and that we will explore in this book. In the next sections we will briefly look at the historical roots of this research area, the textbook content and some considerations in HAI research.

1.2 History of Human–Animal Interaction Research

In its widest possible definition, the study of HAI investigates the influence of non-human animals on humans, as well as our influence on them. Although humans have shared their environment with other animals forever, and humans have domesticated animals for thousands of years (e.g. estimations are that dog domestication happened around 30,000 years ago (Chapter 9, this volume) and farm animal domestication around 10,000 years ago (Chapter 10, this volume)), the history of HAI research is much more recent. The study of HAI uses approaches from social and life sciences (natural sciences), such as veterinary science, medicine, biology, psychology and anthropology, but also from the humanities, such as literature, history, religion and art. Therefore, the multidisciplinary or interdisciplinary area of HAI research is built on knowledge and methodologies developed by many different disciplines (see Chapter 2, this volume). Until

DOI: 10.4324/9781003221753-1

recently, these were applied in isolation, limiting our understanding of the intricate interactions between humans and other animals. In addition, it is thought that the relative recency of HAI research may have arisen, in part, because of lack of interest in human–animal research in some disciplines such as psychology (Amiot & Bastian, 2015) and in part, because of historical theoretical disagreements between disciplines (see example between animal psychologists and ethologists (study of animal behaviour): Dewsbury, 1992). The interdisciplinary study of HAI has mainly developed since the 1980s which initially focussed on understanding the bond between humans and companion animals (Hosey & Melfi, 2014). HAI research subsequently included farm animals, with an emphasis on animal welfare and productivity. Later on, interdisciplinary HAI research was used to explore human interactions with wild animal populations, laboratory animals and zoo animals.

As the field encompasses multiple disciplines, there is no single agreed-upon term or concept that captures how to study or research issues in this area. However, common terms used in life sciences are human–animal interaction (146,000 hits on Google – October 2022) or 'Anthrozoology' (121,000 hits on Google – October 2022), derived from the Greek concepts *anthropos* (human) and *zoion* (animal). Anthrozoology is often defined as 'the scientific study of human–animal interaction' (DeMello, 2012). Other terms are principally used in humanities-related disciplines such as human–animal studies (341,000 hits on Google – October 2022), or human–non-human studies (7,000 hits on Google – October 2022). Human–animal studies is often defined as 'exploring the spaces that animals occupy in human social and cultural worlds and the interactions humans have with them' (DeMello, 2012). All these terms are not equivalent, nor can they be used interchangeably as their objectives are not the same (Collado et al., 2023). Each of these terms is associated with courses or journals that have these names or variations of them in their title (see www.animalsandsociety.org/resources/resources-for-scholars/human-animal-studies-journals/ for a list of human–animal journals). For example, the journal *Anthrozoös*, founded in 1987, describes itself as 'a multidisciplinary journal of the interactions between people and other animals' and is the journal of the International Society of Anthrozoology. HAI courses can be found in many countries across many different academic subjects (see www.animalsandsociety.org/resources/resources-for-students/degree-programs/ for a list of HAI courses). A recent paper was also published offering some guidelines for navigating the different pathways for careers in HAI (Erdman et al., 2023), which might be helpful for the reader. In this book, we use the term 'human–animal interaction' to cover as wide an area as possible while respecting cross-discipline terminology. Our focus is on contributions by social and life sciences.

1.3 Why This Textbook?

Our textbook focuses on the human dimension of interacting with other animals, integrating both social and life or natural sciences approaches. While this book has an anthropocentric approach, the animal perspective is always present. Our textbook covers classic and recent developments, and theories and debates in HAI. Although a humanities-related perspective (e.g. history, art, literature) is interesting (DeMello, 2012), this falls outside the scope of this book. This volume also highlights bias in the field and differences in the methodologies used to study diverse types of HAIs including companion, farm and wild animals. Therefore, our textbook will bring essential knowledge and skills to people who are first introduced to HAI research, but it will also initiate discussion and develop critical analysis, essential for expanding future research directions. The book is not only suitable for students, but also policymakers, professionals such as vets, and NGOs interested in this research area. This textbook consists of a compilation of book chapters

Figure 1.1 QR code of this textbook website https://human-animal-interactions.blogs.lincoln.ac.uk/.

written by experts in their respective areas (see details on each chapter below), and practical pedagogical materials which facilitate both learning and teaching. Covering the entire HAI area would not be possible in an introductory textbook. However, our book aims to be representative of the main HAI topics and as encompassing as possible. In addition, where we feel the reader could benefit from supplementary information, we have provided links to web resources on HAI where readers can find some recommended reading and conference presentations (website link Figure 1.1)

Ten chapters follow this chapter to provide an overview of the different areas of HAI. These chapters are written by experts from both social and life science disciplines, including psychology, anthropology, biology, criminology, ethology, veterinary science and cognitive neuroscience.

Chapter 2 explores the main theories in HAI to provide the reader with different perspectives to contextualise the information presented in this book. This chapter is followed by Chapter 3 on human–animal communication. Since communication takes place in any human–animal interaction, it is essential to comprehend what the key features of human and of animal communication are, and what the biological constraints are when humans and animals communicate.

After investigating the theories and mechanisms underpinning our interactions with animals, subsequent chapters explore the implications of these interactions for humans and animals. Chapter 4 on bioethics introduces the readers to key ethical concepts around our treatment of animals. This leads to Chapter 5, which considers human–animal welfare, stressing how human and animal welfare influence each other. As our attitudes and behaviours towards animals are not universal, Chapter 6 provides some insights into the cultural differences in HAI, through three main examples: human–dog, human–equine and human–chicken interactions. Chapter 7 explores the legislation around our treatment of animals. This chapter provides the reader with an overview of HAI broadly related to forensic contexts, including wildlife crime, animal abuse and the use of animals in forensic science. Furthermore, Chapter 8 gives an outline of different types of animal-assisted interventions, including their definitions, their history and purpose. This chapter offers the reader some critical points and recommendations for the future of this rapidly growing field.

Finally, the three last chapters provide an overview of current research into specific HAIs (i.e. human–pet, human–livestock and human–wildlife interactions). These chapters give the reader an opportunity to look critically at the different approaches used to study each type of interaction. Chapter 9 highlights how human–pet interaction research is predominantly interested in understanding the bond between humans and their companion animals. Chapter 10, which focuses on human–livestock interaction, considers the association between animal welfare and productivity, but is at the forefront of debate in also looking at the interaction with human welfare. Lastly, Chapter 11 presents human–wildlife research which aims to understand human attitudes and beliefs towards wildlife to provide insights into conflicts about wildlife. It also discusses the growing interest in exploring the use of wildlife in entertainment (e.g. circus, film, zoo, wildlife tourism).

1.4 Considerations in Human–Animal Interaction Research

The pedagogical aim of this textbook is to elicit the reader's critical thinking about the rapidly growing HAI research area. Therefore, we would like to highlight some general considerations applicable to the different topics presented in this book. We feel that it is important to understand the points made below for navigating the rich literature in this field, and possibly for stimulating the development of different approaches to study HAI.

1.4.1 Terminology Matters

The terminology used in HAI research can be confusing because sometimes different terms are used interchangeably by different authors, even though the meanings are slightly different. This results from HAI research being multidisciplinary by nature (see Section 1.2). For example, the terms HAI, human–animal relationship (HAR), and human–animal bond (HAB) are often used interchangeably in HAI research, despite some differences in their definitions (Hosey & Melfi, 2014; see Box 1.1).

Box 1.1 HAI, HAR, HAB Definitions

Human–Animal Interaction (HAI)

HAI is defined as an event between two individuals, one an animal, the other a human (Hosey & Melfi, 2019). This event can be a unidirectional action or behaviour from one individual to the other, or a bidirectional action when both individuals react to each other. HAI is the basic level of interaction from which both HAR and HAB derive.

Human–Animal Relationship (HAR)

Repeated HAI results in human–animal relationship (HAR), which comprises past interactions, present ones and predicted sequences of future ones (i.e. anticipating what the other is likely to do; Rault et al., 2020).

Human–Animal Bond (HAB)

With repeated positive HAIs, relationship quality can improve over time, which can lead to the development of a human–animal bond (HAB). A HAB is therefore defined as (1) a relationship between two individuals, a human and a non-human animal (2) that is reciprocal and persistent and (3) promotes a feeling of well-being in both parties (Russow, 2002).

In addition, some terms are so popularised that their initial meaning is lost. For example, the term 'ecotourism' was set as a gold standard for wildlife tourism to make sure that the positive contributions to biodiversity/conservation, tourism, and local communities outweigh the negative impacts of tourism (Ceballos-Lascurain, 1996). However, the term 'ecotourism' has been popularised by the tourism industry (to claim that a product is environmentally friendly as part of 'greenwashing'), and it is now even used in HAI research to refer to wildlife tourism, whether beneficial or not.

Some terms used in HAI research have raised debates around their interpretations. For example, the term 'human–animal interaction' can be interpreted as humans being distinct from other animals, reinforcing the notion of speciesism (i.e. the assumption of human superiority over other animals). The term 'interaction between humans and other animals' would be more accurate, but for simplification, 'human–animal interaction' remains the most popular term (Hosey & Melfi, 2019). Other terms are also debated such as 'companion animal' and 'pet', which are often used interchangeably in research. The term 'pet' is often associated with ownership, reflecting mastership over animals, while 'companion animal' is associated with the psychological bond and mutual relationships (Walsh, 2009).

1.4.2 Ethics

Ethics is paramount to the entire area of HAI. Therefore, throughout the book, ethical considerations are mentioned, apart from having a chapter dedicated to bioethics (Chapter 4).

Academic research is regulated through international and national laws as well as ethical guidelines, which vary between countries (Olsson et al., 2022). Therefore, most academic research must be reviewed by a governance and ethics committee to ensure that the research meets the ethical standards of the institution leading the research before starting any data collection. Such processes are crucial because they serve as a protection/mitigation for the research participants (humans and animals) as well as the researchers. In addition, beyond the philosophical, moral and legislative debates in human and animal ethics, the quality of the research benefits greatly from including ethical considerations into its design, analysis and discussion (Sueur et al., 2023).

Despite the global recognition that assessing the ethical implications in HAI research is important (Olsson et al., 2022), its application is quite disparate. For instance, while human research is often guided by overarching international ethical principles as defined in the 1964 Helsinki Declaration, to date, there is no consensus on international guidelines for animal research (Petkov et al., 2022). Therefore, as human and animal ethics committees are governed by different sets of legislation and guidelines, these committees are often operated separately within an institution, which can lead to poorly informed ethical review (Olsson et al., 2022). This is a particular challenge for HAI research because both human and animal ethical welfare must be considered. Thus, the field would benefit greatly if these dual ethical challenges that HAI research presents were taught in HAI courses and ethics committees.

We provide additional resources about ethical guidelines if the reader would like to explore the topic further (available through our website Figure 1.1)

1.4.3 Potential Research Bias

In this section, we highlight some potential biases or prejudices within HAI research to enhance the reader's critical analysis when reading this book and any HAI research, as well as when developing their own research.

1.4.3.1 Cultural and Gender Representation

As stressed throughout this book, culture and gender can influence how we perceive and interact with animals (e.g. Chapters 6 and 7, this volume). Therefore, it has been suggested that the lack of cultural and gender diversity could impact on theoretical assumptions in HAI and in HAI research (e.g. prioritising some research questions, designs, sampling populations and data interpretations). For example, Herzog (2021) highlighted that a lack of cultural and gender diversity has raised some concerns in human–animal bond research. Most research into human–dog interactions is based on people who are 'WEIRD' (Western, Educated, Industrialised, Rich and Democratic countries), and on dogs considered 'NATIVE' (i.e. Neutered, Alimented (fed), Trained, Isolated, Vaccinated and Engineered (bred)). However, these interactions are not representative of most human–dog interactions in developing countries: 85% of dogs run free (Koster, 2021). Herzog (2021) also showed that out of 64 studies published in the journal *Anthrozoös* between 2019 and 2020, 43 studies mostly used women as participants (over 60% of the sample size), and only a handful of studies included a non-binary sample. Therefore, this raises concerns about the generalisation of these findings because existing participant samples may not be representative of the population.

Overall, it is important that cultural and gender diversity be represented in HAI research. Unfortunately, to date, there is limited data on cultural and gender representation within the HAI research area outside of HAB, especially in relation to the LGBTQQIP2SAA+ communities and other groups than 'WEIRD'. Here, we provide some useful links to resources that might inspire HAI researchers (available through our website Figure 1.1).

1.4.3.2 Species Prejudices

Not all animal species are equally studied in HAI research (see Chapters 2 and 10, this volume). Companion animals such as dogs are studied most, representing over 41% of the 1,715 research papers published between 1998 and 2018 in peer-reviewed journals (Yatcilla, 2021). Other species have rarely been researched, creating a clear gap in our knowledge about HAIs. For example, research into human–fish or human–insect interactions represents less than 1% of the research published between 1998 and 2018 (Yatcilla, 2021). Therefore, we encourage the reader to explore the diversity of interactions humans might have with different species, and we have provided some useful resources here (available through our website Figure 1.1).

1.4.4 One Biology, One Health, One Welfare

Tarazona et al. (2020) make the case that global challenges for humanity can only be addressed by looking at the intersection between animals, humans and the social and physical environment we live in (Figure 1.2). They show that such global challenges as, for example, climate change and disease are influenced by, but also must be addressed by, taking all three elements into account.

Tarazona et al. (2020) make the point that, for example, human-caused climate change has an impact on animals, on humans and on the environment we live in. However, they also argue that solutions in relation to C02 management impact on domesticated animals and our environment. For example, if we choose to eat less meat to curb C02 reduction, this will lead to different crops being grown and fewer farm animals being kept. These choices then lead to new welfare issues for domesticated animals, for example, keeping smaller numbers of animals under conditions that generate less C02 ('sustainable production'). Another example of the interconnection between humans, other animals and our shared environments is the zoonotic (animal based) disease Covid-19 or SARS-CoV-2. Human infection with SARS-CoV-2 led to self-isolation in many countries, resulting in a change in the physical environment (e.g. less C02 emissions for a limited period of time; Weir et al., 2021) as well as important environmental changes for animals (e.g. there was an impact on habitat quality in urban areas in different ways for different species (Coman et al., 2022) as well as fewer roadkills (Driessen, 2021)). These examples show that none of the elements in Figure 1.2 can be studied in isolation.

The concept of One Biology, One Health, One Welfare gives an aspirational perspective to HAI: that we should strive to study humans, animals and our environment in interaction, since we have a shared biology (e.g. animals and humans are subject to the same evolutionary principles), and due to interacting in the same environment, we can only understand our health and our welfare in relation to each other. Not all authors use the three terms together as Tarazona et al. (2020) and as we have done here. For example, Pinillos et al. (2016) and Stephens (2021) focus on One Welfare, whereas Mackenzie and Jeggo (2019) and Rushton and Bruce (2017) focus on One Health, with Lindenmayer and Kaufman (2021) talking about One Health and One Welfare as well as Leconstant and Spitz (2022) (see also Chapter 5, this volume).

The One Biology, One Health, One Welfare approach points to the need for multidisciplinary or interdisciplinary cooperation. For example, in their review article about the link between animal cruelty and interpersonal violence, Mota-Rojas et al. (2022) point out that cooperation between dentists (damage to the mouth can be an indicator of physical abuse), and veterinarians (signs of animal abuse) can lead to detecting domestic abuse. However, they also argue that interdisciplinary cooperation can play a role in mitigating against the effects of abuse or play a role in preventing it: for example, educational programmes that include animal-assisted activities for prisoners and those who are the victims of abuse can either assist in developing empathy and kindness or can give victims the empathy and kindness that is needed.

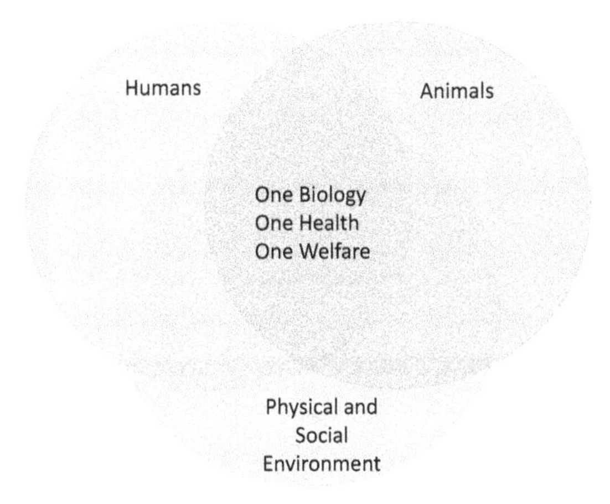

Figure 1.2 The intersection between humans, animals and the physical environment: One Biology, One Health, One Welfare.

1.5 Conclusion

The research area of HAI is relatively new and based on multiple disciplines. This means that the terminology used in this area is sometimes confusing. In this chapter, we have tried to clarify some of the terminological pitfalls, as well as to provide some perspectives to study the HAI area critically and constructively. We have highlighted some areas that need attention when working or conducting research in this field (ethics, and cultural as well as gender and species biases). Finally, we have shown the importance of not studying HAI in isolation from a shared environment.

2 Theories in Human–Animal Interaction

Emile van der Zee

2.1 Introduction

Theories in human–animal interaction (HAI) derive from some academic subjects that stood at the cradle of this new field, for example, psychology and biology. This means that some of the theories relevant in HAI may originally have been designed for different purposes and adapted to this area of research (such as attachment theory in psychology, which was originally designed to explain carer–child relationships) or have only seen the limelight very recently (such as 'the animal connection' hypothesis which stems from palaeoanthropology). Consequently, predictions about our complex, ambiguous and ubiquitous relationship with other animals sometimes overlap, or venture into different HAI areas.

As with many theory overviews in the area of HAI, this chapter will mostly follow Beck's (2014) overview of theories, although updated and expanded. The first section will look at how some theories try to explain our interactions with other animals and will give in each case a few examples where these theories have been or can be tested. It should be noted that many more theories and models exist in HAI. Highly specialised theories, such as those developed for the use of animals in school settings, are not considered (e.g. Gee, Griffin & McCardle, 2017). This chapter will only consider general theories that allow us to understand the materials presented in this book. Subsequent sections will also look at how anthropomorphism (assigning human properties to non-human entities) may stand in the way of understanding HAI.

2.2 The Biophilia Hypothesis

The social psychologist and psychoanalyst Erich Fromm coined 'biophilia' as 'the passionate love of life and of all that is alive' (Fromm, 1973, p. 365). He saw biophilia as a biological impulse, opposite to the human tendency to destroy, to be aggressive. The socio-biologist E. O. Wilson (1984) grounded this biological impulse in a genetic background. In his book *Biophilia*, Wilson (1984) argues that humans are genetically determined to affiliate with other living systems. Furthermore, it is argued in Kellert and Wilson's (1993) book *The Biophilia Hypothesis* that this 'inborn need for nature' (p. 448) even drives us to use technology to create life, and search for life. According to Wilson (1984), our frustration of losing touch with nature in a modern world, where nature is often remote, can be overcome by environmental stewardship: for example, by being involved in the conservation of species, and the preservation of our natural surroundings. But what is the evidence for the biophilia hypothesis?

Chang et al. (2020) looked at 31,534 photographs posted on Flickr across 185 countries over a period of 11 years. The authors found that photographs of water, terrestrial landscapes, plants, animals and nature were associated more with fun activities, weddings, celebrations,

DOI: 10.4324/9781003221753-2

honeymoons and vacations compared to regular daily activities (used as a baseline measure). In addition, countries with more plant-related nature in photographs depicting fun activity had a higher life satisfaction (e.g. Costa Rica and Finland). Tam (2013) considered many different questionnaires that measure connectedness to nature (e.g. commitment to nature, emotional affinity towards nature, inclusion of nature in self), and discovered that – as predicted by the biophilia hypothesis – it is possible to identify 'connectedness to nature' as a single factor underlaying the measurements in all of these questionnaires. In addition, Tam (2013) found that scoring high on the different questionnaires correlated positively with an increase in subjective well-being. People tended to associate connectedness to nature with life satisfaction (a cognitive component), and a feeling of pleasantness (an affective component).

The therapeutic and restorative effects of spending time in nature have been documented widely (e.g. see Chavaly & Naachimuthu, 2020 for an overview). However, is there a specific benefit in interacting with animals as well? Thayer and Stevens (2022) provided evidence that we profit from an interaction with animals when it comes to affect and stress, but possibly not when it comes to specific cognitive tasks. After being given the opportunity to sit calmly with an unfamiliar dog for three minutes, followed by interacting with an unfamiliar dog for around 40 minutes, participants filled in several questionnaires and performed several tasks. Compared to control conditions where participants did not sit calmly or interacted with an unfamiliar dog, positive affect and a reduction of anxiety and stress were observed. However, long-term memory, working memory and attentional control (all executive functions) did not improve. This means that there are benefits in our interactions with animals, but that we must also be aware of the limits of these advantages in our psychological functioning. Chapters 8 and 9 (this volume) also show that evidence for the potential benefits of interacting with animals is mixed.

According to Kellert and Wilson (1993), biophilia is inborn, but it is not an instinct: a fixed principle which, once activated, fully determines our behaviour. We can also be biophobic: have a fear of animals, plants or organic materials. The diagnostic criteria for a phobia are that the phobic object or situation almost always provokes immediate fear or anxiety, that the object or situation is avoided or endured with intense fear or anxiety, and that the fear, anxiety or avoidance is persistent, typically lasting for six months or more (American Psychiatric Association, 2013). The list of biophobias is long, and often refers to highly specific instances of things that are alive. To name a few: botanophobia (fear of plants), anthophobia (fear of flowers) and ornitophobia (fear of birds). Some phobias may have an innate component. For example, six-month-old infants react with arousal, as measured by their pupillary response, to spiders and snakes, but not to flowers and fish (Hoehl et al., 2017), suggesting a predisposition to fear spiders and snakes. Therefore, biophilia is part of a continuum, with biophilia itself at one extreme end, and biophobia on the opposite end. Specific animals can be located at different locations on the continuum at the same time for one individual (Figure 2.1). While the biophilia hypothesis is intended to explain our ubiquitous need for interaction with nature and also animals, biophobia may explain our fear or even hate of nature and of non-human animals, with biophilia and biophobia together intending to explain our complex and sometimes ambiguous relationship with nature and non-human animals.

While there is some support for the biophilia hypothesis in relation to the human–nature relationship, evidence in the HAI domain is less clear. For instance, Joye and De Block (2011) argue that the biophilic notion and the evolutionary reasoning behind the hypothesis are not clear. In addition, the evidence is mostly based on self-report measures, which measure what we believe to be true about ourselves, but do not measure how we behave. See also Chapters 8 and 9 (this volume).

Biophilia Biophobia

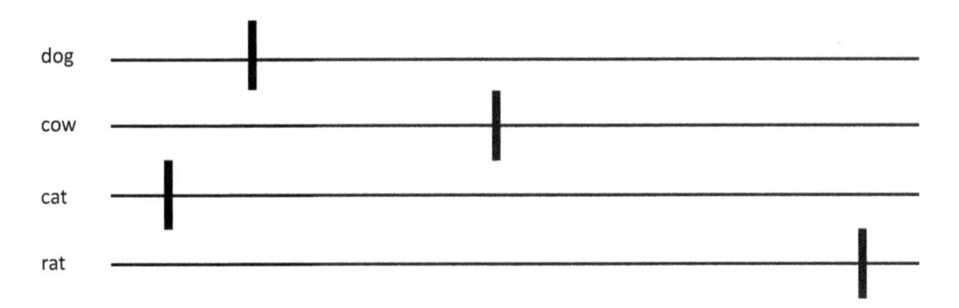

Figure 2.1 The biophilia–biophobia continuum, where specific forms of life can be located at extreme sides at the same time for one individual. Vertical bars represent 'sliders' on the continuum for a particular animal. Here we see a hypothetical individual who likes cats more than dogs, is indifferent to cows, but has an aversion to rats.

2.3 Neoteny Preference

Neoteny is the phenomenon of adults retaining traits of juveniles of the same species. Not only young but also adult humans are neotenous (see Kollman, 1905; Shea (1989); Bufill et al. (2011) for a wider debate). Adult humans have, for example, large heads compared to body size, flat and broad faces, small teeth, small upper and lower jaws, (mostly) hairless bodies and short legs in relation to their torso; exactly what one would expect to see in a juvenile human being. In addition, we appear to prefer juvenile traits in our own species. As observed by Beck (2014) human-like cartoon figures have, over the years, evolved to have larger eyes, larger heads and smaller bodies (e.g. see Micky Mouse, Paw Patrol and My Little Pony). The preference of neotenic traits in our own species even impacts on mate selection: across different cultures, adult human males prefer adult females with neotenic facial features: small noses and ears, and full lips (Jones et al., 1995).

Our behavioural responses to neotenic features in our own species are quite specific: the human response to babies is approaching, helping, using baby-talk and caring (Glocker et al., 2009; Zebrowitz et al., 1992). From a very young age this so-called baby scheme response can not only be seen in interaction with humans but also other species (Borgi et al., 2014). For example, we use baby talk or infant-directed speech to talk to both younger and older dogs (Ben-Aderet et al., 2017), showing that the age of a member of another species is not a determining factor.

Could it be that our relationship with animals is based on this neoteny preference but applied to animals that show similar neoteny features? Several studies suggest that this is the case. For example, Kruger (2015) showed that we are more likely to want to hold animals, pet them and are willing to adopt them if they have neotenic features that remind us of human babies. In his study, Kruger (2015) looked at non-mammalian animals that require parental care, such as birds like the emperor penguin and reptiles like the dwarf crocodile. He demonstrated that these animals are perceived to have more neotenic features (cute, immature, helpless) than animals that do not require parental care (birds like the black-headed duck and reptiles like the Californian alligator lizard). He found that, although women were more likely to adopt non-mammalian infants that require parental care than men, both men and women wanted to hold these animals, as well as pet them. This suggests that neoteny preference may be a driving force in our interaction with other species, even in relation to reptiles.

Domestication (selective breeding) may increase the number of neoteny features in some animals (Meehan & Shackelford, 2021; Chapter 9, this volume). This 'domestication syndrome' leads to increased tameness, supernumeracy toes, depigmentation and curly hair, as well as decreased brain sizes, shorter muzzles, floppy ears and so on in dogs, cats, pigs and cattle. The latter features increase the cuteness factor in these animals. Although not all domestication leads to the domestication syndrome (Sánchez-Villagra et al., 2016), it is important to be aware of our own role in breeding animals that look cute: breeding cuteness may lead to animal welfare issues (Serpell, 2003). Diseases in the dog as a result of selective breeding are good examples: coat colour-related sensorineural deafness (Webb & Cullen, 2010), Alzheimer's disease (De Risio et al., 2015), epilepsy (Dewey et al., 2019) and an obstruction of the airways as a result of breeding flat faced (brachycephalic) dogs like the bulldog (Dupré & Heidenreich, 2016). Human neoteny preference can thus explain our ubiquitous interaction with certain species, but also the motivation of wanting to care but failing to do this properly.

2.4 The Animal Connection

Shipman (2010, 2011, 2015, 2021) takes Wilson's idea of a genetic basis for human–animal interaction a step further. She firstly establishes that humans are unique in being genetically predisposed to interact with other animals. For example, she observes that we are unique in cross-species alloparenting: we take care of (the young of) other species as if they were our own offspring. The few cases where other animals take care of another species – and truly feed and protect for a sustained period – are extremely rare (e.g. the parasitic cuckoo catfish which manages to have its fertilised eggs be cared for by mouth-breeding fish is the only example of non-avian brood parasites among vertebrates; Zimmermann et al., 2022). However, more importantly, she argues that our universal inclination for an animal connection has influenced our own evolution, genetics and behaviour. She identifies three human adaptations, that are caused by and accompanied by our increased focus on the animal connection: tool making, symbol use (language and art), and the domestication of animals and plants.

Shipman argues that the ability of some of our ancestors to predict prey and predator behaviour gave them an evolutionary advantage: fewer injuries and more energy for reproduction. Those with an animal connection thus had an evolutionary advantage that was passed on through successful survivors in our species. Stone tools that were initially developed for processing carcasses allowed us to process more meat, and at a faster rate than before. Our reading of animal behaviour and our invention of stone tools together made us a top predator, despite a relatively weak bodily constitution.

Shipman points to the fact that early cave paintings mostly contain animals of an intermediate size (the ones we interact with), and rarely small animals, or even humans. And she argues that our domestication of animals well before the end of the last ice age (which ended around 10,000 years ago) points to our use of animals as an extension of tool making: the domestication of the wolf around 40,000 years ago, for example, helped us hunt, in a near symbiotic relationship, where both humans and dogs profited from more efficient hunting techniques, giving both species access to more meat. She even speculates that it was the domestication of the dog that gave modern humans an advantage over our Neanderthal cousins (Shipman, 2015). Further detail about pet domestication is available in Chapter 9 (this volume).

According to Shipman, domestication changed the genetics and the food intake as well as our behaviour and that of domesticated species. For example, humans developed genes for breaking down milk (lactase) that we receive from cattle, and dogs developed genes for breaking down wheat (amylase), which is an advantage when sharing our food with them (Arendt et al., 2014; Axelsson et al., 2013; Shipman, 2015).

Shipman is a paleoanthropologist, and the predictions of her theory do not focus on present-day human–animal interactions. She does, however, predict physiological and behavioural changes that must have taken place in the past, and that can be independently investigated. She predicts, for example, (1) that remains of domesticated wolves will not be found earlier than the arrival of early humans in Eurasia, and will not be found at Neanderthal sites, (2) that sites that do not have domesticated wolf remains will be smaller and will contain fewer bones of prey animals, (3) that sites that have domesticated wolf remains will have remains of larger prey animals (e.g. mammoths), due to the combined ability of hunting efficiently, and (4) that around such sites remains of non-domesticated wolves may be found that were killed by humans (who protected their domesticated wolves). So far, no domesticated wolf bones have been found in Neanderthal sites, but have been found in sites of modern humans, confirming and not contradicting one of her predictions. Even if some of Shipman's other predictions have not yet been tested, the animal connection hypothesis is rich in describing the set of possible reasons for our ubiquitous relationship with other animals, and it can be seen as a further step to explain the genetic basis for our love of certain animals. It offers an interesting set of genetic and behavioural changes we and the animals we have domesticated went through, and it provides a possible insight into our ambiguous relationship with canids.

2.5 Attachment Theory, Social Support Theory and the Biopsychosocial Model

Attachment theory (Ainsworth et al., 1978; Bowlby, 1969; Main & Solomon, 1986) and social support theory (Cohen & Wills, 1985) originally stem from psychology, whereas the biopsychosocial model originates in medicine (Engel, 1981). These theories are discussed together here because HAI versions of these theories focus on the human–animal bond (HAB), and the predictions of these theories also overlap.

According to attachment theory (Ainsworth et al., 1978; Bowlby, 1969; Main & Solomon, 1986), the emotional bond between a carer and a child can be any of four types: secure, anxious, avoidant and disorganised (Table 2.1). In most cases, children develop a secure attachment to their carers: avoiding a stranger when they approach and seeking comfort if a carer is present to reduce tension and anxiety. A child is insecurely attached if it becomes anxious or avoidant towards the carer. An anxious child depends on their carer, is distressed when the carer leaves and may act out. An avoidant child may ignore their carer if a stranger approaches, showing reliance on themselves. Disorganised attachment is when a child freezes or shows signs of confusion; the child can be passive or unresponsive. All four types are said to be universal, can be determined by the way a child acts if a stranger approaches, and are based on different levels of trust in oneself and the carer.

Table 2.1 Different types of carer–child attachment

	High Trust in Self	*Low Trust in Self*
High Trust in Carer	**Secure**	**Anxious**
Low Trust in Carer	**Avoidant**	**Disorganised**

Cohen and Wills (1985) discussed the beneficial effects on dealing with stress by receiving direct social support from other humans, and the protection that social support may offer against the negative effects of stress or threatening events. As observed by Beck (2014), this so-called stress buffering, and also attachment, not only applies to humans but also to pets and other non-human animals.

Using human attachment-theory, Zilcha-Mano et al. (2012) observed that dog and cat owners who were securely attached to their pets, and with their pets present or with descriptions the owners made of their pets being present, generated more life goals compared to owners with anxious or avoidant attachment. These authors also found that pet presence (physically or through a self-generated description) had a lowering effect on the blood pressure of securely attached owners during a stressful task. Such an effect was not found for owners with an avoidant or anxious attachment. Importantly, pet attachment was found to be independent of human attachment. The authors interpret their results as showing that pets provide a uniquely safe haven for exploration and growth, and for helping in stress reduction. This study thus provides evidence for stress buffering as proposed by social support theory, as well as attachment theory in explaining human–pet interaction. Furthermore, it also refers to a biological component (blood pressure) which mediates the positive effect of companion animals on human well-being (Chapter 9, this volume).

Meehan et al. (2017) show that companion animal owners regard dogs and cats as a source of social support, although to a slightly lesser extent than significant others, family and friends. Looking more in depth at attachment components, these authors found that proximity seeking is higher towards pets than towards mothers, fathers and siblings, although lower compared to significant others and friends. When it comes to providing a safe haven, being distressed when separated, and providing a secure base, pets were deemed more important than siblings, although less important than other social relationships. This study thus provides evidence about our complex attachment relationship with pets.

Other typologies than those based on social support theory or attachment theory that characterise our interactions with pets are available in the literature. For example, Blouin (2013) distinguishes between three different types of pet owners: (1) dominionistic: owners who see their pets instrumentally, for example, the protection they provide, (2) humanistic: owners who see their pets as surrogate humans and focus on the affective nature of the relationship, and (3) protectionistic: owners who see their pets as companions, and as autonomous beings. This typology provides an additional characterisation of the different relationships we can have with our pets (Blouin, 2012; Chapter 9, this volume).

The biopsychosocial model offers a comprehensive view of the possible psychological and social factors that, in combination with biological factors can impact – positively or negatively – on our health when interacting with companion animals (Friedmann & Gee, 2017). Polheber and Matchock (2014) exposed participants to a stressful situation (the Trier Social Stress Test). Participants were asked to do a five-minute interview for a job they were asked to imagine wanting to have, with only three minutes of preparation. The preparation and interview were done in front of a panel of three people, with the entire process recorded on audio and video, and with panel members being mostly apathetic and unresponsive. This was followed by a difficult mental math test. Participants were either accompanied by a friend, by a dog they did not know, or were not accompanied (control condition). Cortisol levels (a stress hormone) and heart rate were measured before the intervention, directly after the intervention, and 23 minutes after the second sample. Compared to the friend and control conditions, cortisol levels and heart rate were lower throughout the entire experiment when participants were accompanied by a dog. Stress measures showed that all participants were more stressed after the experiment, compared to before the experiment, showing that the stress manipulation worked. This experiment thus indicated that there was an attenuated physiological stress response if participants were accompanied by a dog, but not by a friend (or unaccompanied). According to Friedmann and Gee (2017), this research illustrates the beneficial health effects of companion animals and underlines the advantages of using the biopsychosocial model when studying human–companion animal interaction. Chapters 8 and 9 in the present volume use the biopsychosocial model when looking at human–pet interaction.

Support from, and attachment to, animals can also have negative dimensions, ranging from overdressing pup dogs as babies to zoophilia – sexual fantasies about and/or the engagement in sexual activities with animals (bestiality: see Holoyda et al., 2018). Beetz (2005a) gives an overview of several studies on bestiality that were all based on self-reporting adults. Although many different types of mammals were mentioned, dogs featured highest on the list, followed by horses and cattle. Seventy-six per cent of participants who engaged in bestiality across all studies reported a very strong emotional attachment to the animals they exploited (comparable to the love for a human partner), and 79% reported having fallen in love with the animal (Beetz, 2002). Miletsky (2002) reports that among participants engaged in bestiality, 100% of women and 91% of men mentioned sexual attraction as the main motivator for having sex with animals. Revealing is that the exploitation was not based on the unavailability of human partners; only 12% of participants reported this as a motivating factor (Miletsky, 2002). The negative welfare effects for the animals concerned can be severe: gross lesions, the presence of foreign bodies in the animal, and abrasions and bruises (Stern & Smith-Blackmore, 2016). What these studies highlight, however, is that although animals in general or pets in particular may provide social and emotional support for us humans, the flip side is that animal welfare may be compromised. See Chapter 7 in this volume for more forensic issues in relation to animals.

Research related to the three theories on human–pet interaction in this section thus also reveals some of the reasons for our ubiquitous and sometimes ambiguous and complex relationships with our pets; we seek their company to feel better, but in doing so we may compromise their welfare – especially if we treat them like other human beings. Section 2.6 further investigates why we sometimes ascribe human traits to other animals.

2.6 Anthropomorphism

We see human faces in the mountains on Mars, and Jesus in the burn marks on a piece of toast. We give human names to our pets, cars, plants, ships and guitars. And, as already observed by Heider and Simmel (1944), we even spontaneously attribute intention and cause of motion to abstract moving figures like triangles and circles. Assigning human properties to objects, events, situations or other things is widespread. We assign human properties on the basis of analogy or homology (similarity in form) to ourselves (Guthrie, 1997).

Brandt and Reyna (2011) noted that anthropomorphism is part of a continuum, where we attribute human characteristics to entities lower in the great chain of being (e.g. objects, plants and animals), or higher levels (e.g. a God or a collection of gods) (Figure 2.2.).

Anthropomorphism must be interpreted in relation to other continua: we can sanctify human beings, animals, plants or objects, thereby elevating them to a god-like status, and we can dehumanise things at a higher level of existence to a lower level of existence. For example, we can objectify animals as tools to satisfy our needs, or we can associate people with cockroaches, rats or other vermin.

What exactly happens, though, when we anthropomorphise? Let us consider the empirical evidence.

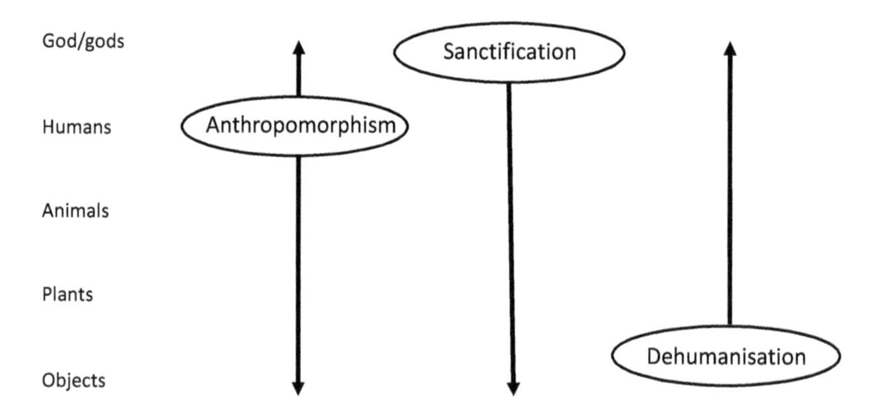

Figure 2.2 The continuums of anthropomorphism, sanctification and dehumanisation. Anthropomorphism projects human characteristics to all levels of being: objects, plants, animals and god/gods. Sanctification projects properties from a higher level to a lower level (making everything god-like), and dehumanisation projects properties from a lower level to a higher level (objectifying everything).

Eddy et al.(1993) investigated how similar we feel to other animals, and what cognitive abilities we assign to other animals (Table 2.2). To determine similarity, participants were asked to indicate on an 11-point Likert scale (from 'not at all similar' or 0 to 'identical' or 10) the degree to which they felt that animals were similar to them, and the degree to which they felt that animals experience the world in a way that is similar to how participants experience the world. To determine cognitive ability, participants were asked on an 11-point Likert scale (from 'not at all possible' or 0 to 'very possible' or 10) whether the animal could trick another animal into going to a place where it knows food is not present, in order for it to get the food itself; whether they felt the animal could figure out that the image it sees in a mirror is itself and not another animal; and whether the animal could distinguish being tripped over or kicked. Their findings can be found in Table 2.2.

Some interesting patterns emerged from these findings: there is a correlation between similarity and cognitive ability ratings. Both kinds of ratings follow the phylogenetic scale (the hierarchy of evolutionary related species). Pets and higher primates are separate categories within the category of other mammals, with large differences between companion animals and farm animals. On the basis of these findings, the authors concluded that anthropomorphism is derived from our ability to infer cognitive mental states in other animals.

However, other explanations as to the foundations of anthropomorphism exist. For example, Harrison and Hall (2010) give priority to genetic relatedness (as in Table 2.2), but point out that this correlates with the perceived ability of animals to have empathy with humans, and the perceived ability of the animals to communicate with us (two measures which themselves correlate well). Phylogenetic rank also predicted the use of the pronouns 'he' and 'she' as opposed to 'it' in English.

An interesting alternative view is that behavioural similarity to humans in a similar context drives anthropomorphism (Hamm & Mitchell, 1997). Hamm and Mitchell (1997) depicted a child, monkey and dog in a context of jealousy, and a human, chimpanzee, bear, elephant, otter and dog in a context of deception (emotional concealment) and show that for these two factors, behavioural similarity to humans predicted anthropomorphism, but not physical similarity,

Table 2.2 Ratings from 0 to 10 on how similar other animals are to us, and to what extent they are perceived to have the same cognitive abilities that we have

Perceived Similarity Category	*Animal*	*Cognitive Index*
Identical (9–10)	Human	9.7
Very similar (7–8)	Chimp	7.5
	Monkey	7.3
	Gorilla	7.1
Moderately similar	Dog	6.6
	Cat	6.1
	Cheetah	5.1
	Porpoise	3.9
	Cow	3.6
	Pig	3.6
	Parrot	3.0
	Eagle	3.5
	Elephant	3.8
Slightly similar (2–3)	Goat	3.5
	Canary	2.6
	Parakeet	2.7
	Crocodile	2.6
	Robin	2.4
	Snake	2.2
	Chicken	2.9
	Turtle	2.0
	Frog	1.9
	Trout	1.5
	Goldfish	1.3
Not at all similar (0–1)	Toad	1.6
	Salamander	1.6
	Crab	1.8
	Guppy	1.1
	Worm	1.0
	Cockroach	1.0

Source: From Eddy et al., 1993.

phylogenetic closeness to humans, familiarity as well as having an affectional bond, or cultural stereotype. Much research in this area still needs to be done.

There are many different consequences of anthropomorphism. For example, there is a link between anthropomorphism and vegetarian attitudes. Bastian et al. (2012) show that people are more willing to eat animals if the animals can be denied mental capacity. Participants rating 32 animals showed that the more edible an animal is deemed to be, the less mental capacity it is seen to have. In addition, if meat eaters had the expectation to eat meat during the study, they were less inclined to attribute mental capacity to animals.

Waytz et al. (2010) show that a positive effect of anthropomorphism is to care and to have concern for the environment. Using the Individual Differences in Anthropomorphism Questionnaire (IDAQ), they show that individuals scoring high on the IDAQ also find it wrong to harm a computer, a motorcycle and rare flowers to save a human life. As predicted by the biophilia hypothesis, those scoring high on the IDAQ were also more in favour of protecting plants and trees.

The positive effects of anthropomorphism extend to protecting animals as well. Butterfield et al. (2012) showed that participants were more inclined to help dogs and humans that were described in anthropomorphic language compared to non-anthropomorphic language. Participants (90% dog owners) shown pictures of dogs and asked to rate them for anthropomorphic traits (e.g. having good humour, being a good listener) were more willing to adopt dogs, and were more supportive of animal rights, animal welfare and vegetarian and vegan attitudes than participants who were asked to rate dogs for non-anthropomorphic traits (e.g. a good sense of smell, listening well to commands). Priming participants with anthropomorphic language thus impacted on favourable welfare attitudes.

We have seen other positive effects of anthropomorphism: it may help us understand and predict animal behaviour (Shipman, 2010, 2011). However, anthropomorphism can also have negative effects. Attributing human properties to animals may impact on their physical and psychological welfare. For example, clothing pets may lead to thermoregulation issues, and assigning guilt to dogs (when they have destroyed items in their owner's absence) may lead to inappropriate appeasement behaviours in the dog (Mota-Rojas et al., 2021). But anthropomorphism, positive or negative, can not only have an impact on animals and on us, but also on how we study human–animal interaction: anthropomorphism can influence our research and our theories. When anthropomorphism stands in the way of understanding how animals really are, we can speak of anthropomorphic prejudice.

2.7 Possible Anthropomorphic Prejudices in Human–Animal Interaction

This section considers several possible sources of anthropomorphic prejudice: Do animals feel pain, like we do? Do they have the same emotions that we have? Are they aware of themselves, as we are? Do they recognise internal states in others, as we typically do? These four possible sources of anthropomorphic prejudice may, by themselves, or in combination, have welfare implications. For example, if our research and our theories would determine that animals experience pain, like we do; can be sad, as we are; and are aware of these things, then we may feel morally obliged to provide conditions for animals that prevent pain and sadness, or reduce it.

There is an important constraint on studying human properties in animals, however, and we need to look at this constraint first. Cheney and Seyfarth (1990) make a strong case for using Occam's razor: to prefer a simpler explanation over a more complex explanation where possible. For example, when a vervet monkey gives a call for 'snake', this results in many other vervet monkeys standing on their hind legs and looking around in the grass. We can choose to attribute knowledge to the caller of the internal state of their listeners along the lines of 'my listeners do not see a snake, so I better make them aware of it'. Or we can assume that callers merely signal their own internal state along the lines of 'I see a snake, I call'. In the latter case, intention and motivation are not necessary for an explanation: the caller simply flags up what they believe to be true about the world, and the receiver has learned to avoid danger based on a sound. Cheney and Seyfarth (1990) point out that 'intention' should only be confirmed if allowed by the experimental set-up and should not be assumed (see also Chapter 3, this volume). We need evidence for higher-level explanations. If we do not have it, we should use lower-level explanations where possible.

2.7.1 Do Animals Have Pain, Like We Do?

According to the International Association for the Study of Pain, pain for humans is 'An unpleasant sensory and emotional experience associated with, or resembling that associated with, actual or potential tissue damage' (IASP, 2022). As observed by Sneddon et al. (2014), animals, like babies, cannot tell us whether they are in pain using language, and we can therefore only learn in an indirect manner whether animals are in pain.

Birch et al. (2021) formulated a set of criteria in relation to animals like crabs, octopuses and lobsters that allow us to infer indirectly whether an animal is in pain. They assigned confidence ratings to their criteria, making a quantitative evaluation of pain perception possible. For example, based on the available scientific evidence for the octopus, Birch et al. (2021) gave very high confidence ratings to four out of eight criteria for pain perception (with high confidence ratings for three other criteria and a medium confidence rating for the remaining criterion):

1. The animal possesses receptors sensitive to noxious stimuli (nociceptors).
2. The animal possesses integrative brain regions capable of integrating information from different sensory sources.
3. The animal shows flexible self-protective behaviour (e.g. wound-tending, guarding, grooming, rubbing) of a type likely to involve representing the bodily location of a noxious stimulus.
4. The animal shows associative learning in which noxious stimuli become associated with neutral stimuli, and/or in which novel ways of avoiding noxious stimuli are learned through reinforcement.

Birch et al.'s (2021) burden of evidence for experiencing pain is mostly based on the centralised integrative processing of sensory information specialised in detecting real or potential tissue damage (congruent with the human definition of pain). In cases where the authors did not have a high confidence in an animal satisfying criteria for pain perception, this was due to a lack of positive evidence, and not due to the animals not satisfying the criteria. This raises the issue of how we should deal with animals for which evidence in relation to experiencing pain is lacking.

In the context of animal sentience, Birch (2017) refers to the precautionary principle: 'Where there are threats of serious, negative animal welfare outcomes, lack of full scientific certainty as to the sentience of the animals in question shall not be used as a reason for postponing cost-effective measures to prevent those outcomes' (p. 3). In other words, the fact that we do not know whether animals feel pain does not mean they do not have pain. Especially if welfare measures are affordable, there is no reason not to implement them, just in case.

Based on their research, Birch et al. (2021) advised, for example, against the practice of removing claws in crabs (a common practice in the fishing industry) before putting these back in the water, where these claws may regrow, and also advised against boiling lobsters alive, and separating the head or abdomen from the thorax to kill them. All of these practices are more than likely to inflict pain for a considerable period of time. They advise relatively quick and low-cost slaughtering practices, such as double spiking for crabs and whole-body splitting for lobsters. Similar low-cost slaughtering practices can be applied in cases where the evidence of whether animals have pain or not is less clear, but where the precautionary principle is applicable, to safeguard the welfare of other non-human animals.

Human prejudice in pain perception is generally based on three principles:

1. Recognising that animals feel pain but that, for example, the need for animal research (Gallup & Beckstead, 1988), cultural practices or for obtaining food take priority over those feelings.
2. That animals lower on the phylogenetic scale have more limited cognitive abilities (Eddy, Gallup & Povinelli, 1993), and are therefore not able to experience pain as we do (dehumanising animals; Table 2.2).

3 Recognising that animals experience the world like us (anthropomorphic prejudice; Figure 2.2), making us blind to their needs and unique constitution. For example, the thermal threshold for nociceptors in mammals is 40 degrees Celsius, whereas the thermal threshold for rainbow trout is around 33 degrees Celsius, and even in tropical zebra fish is 36.5 degrees Celsius (Sneddon, 2019). Therefore, a nice hot bath for us may feel like scalding water to a rainbow trout. Only systematic research can remove prejudice, and quantitative research with objective criteria is an important way to do this in relation to pain perception.

2.7.2 Do Animals Have Emotions, Like We Do?

Humans experience a universal set of basic emotions. Apart from our ability to express and comprehend these emotions via language – something animals cannot do as we do (Chapter 3, this volume) – these emotions can be readily expressed in a universal fashion by our faces, and read from them (Ekman & Friesen, 1971): happiness, sadness, fear, surprise, anger and disgust. Although the original research focussed on faces, it is now accepted that these emotions can also be expressed and read from other non-verbal sources, such as body posture (Pollux et al., 2019), and are correlated with specific brain activation patterns (Ekman & Cordaro, 2011). In humans, emotion processing is automatic and gives us an indication of each other's welfare.

Do animals have the same basic emotions that we have? Dogs show distinct facial expressions for positive anticipation and frustration (Bremhorst et al., 2019), and they use different facial expressions for fear and happiness than we do (Caeiro et al., 2017). Laterality in animal behaviour also seems to be associated with their emotional states. Positive anticipation (owner approach) is associated with a right tail wag in dogs, whereas a threat is associated with a left tail wag (a dominant or unfamiliar dog approaching), and when feeding, dogs react with a left head turn for potentially threatening stimuli (seeing a silhouette of another dog, a cat or a snake) (Siniscalchi et al., 2021). We can thus not evaluate a dog's happiness or threat anticipation or fear by assuming it has the same facial or body expressions as we have.

This is not different in relation to other species. Investigating positive valence (anticipation of a food reward) and negative valence (food visible but not accessible) in goats, Baciadonna et al. (2020) found different behavioural and physiological responses for positive compared to negative events: ears backwards more, more active, more vocalisations, and a higher heart rate. Using similar research paradigms Reimert et al. (2017) found different behavioural and physiological correlates for positive compared to negative events in pigs, and Reefmann et al.(2009) found similar differences for sheep. As observed by Machado and da Silva (2020), however, the positions of the ears and tails may vary between species for similar emotional states and there is therefore no common pattern in non-human animals. For example, positive valence is associated with ears going backwards in goats but is associated with ears being upright in the dairy cow (Lambert & Carder, 2019). So, we have to be careful to generalise from one species to another one.

Humans thus have to be careful when trying to associate emotional states with behaviours (facial expressions or gestures) in other animals. We cannot use our own behavioural correlates, which we use when recognising emotions in other humans. Since we cannot ask other animals about their emotional states, our methods to investigate this issue are limited to behavioural (e.g. ear position, tail position, facial expression) and physiological methods (e.g. heartbeat and cortisol measurements). We depend on trained experts supported by research, to overcome our species-specific automatic emotion processing when evaluating other animals' emotions.

2.7.3 Do Animals Have Self-Awareness, Like We Do?

Gallup (1970) developed the mirror test, to test whether chimpanzees have self-awareness (sometimes referred to as 'self-recognition', 'a concept of self' or 'self-consciousness', although see, e.g. Mitchell (1997) for a criticism of this idea). Gallup (1970) exposed four socialised chimpanzees to a mirror for ten days. During that period, the number of socially directed responses (as if seeing another chimpanzee) went down, and the number of self-directed responses went up, as did the total amount of mirror viewing time. Each chimpanzee was then anaesthetised, and the uppermost eyebrow area as well as the top half of the opposite ear were painted red with an odourless non-irritant paint. Then the number of touches without mirror presence were recorded. When the mirror was subsequently introduced, the viewing time doubled and mark-related touches increased 25 times. In socially isolated chimpanzees, self-awareness behaviour was not evidenced (Gallup et al., 1971), and socially isolated chimpanzees that were socialised did show self-awareness (Hill et al., 1970), demonstrating that socialisation (and thus having seen other chimpanzees) was a determining factor.

Human toddlers pass the mirror test from around 18 months old (Nielsen et al., 2006). Other animals that passed the mirror test are other great apes: bonobos, gorillas and orangutans (Gallup et al., 2002). Animals that did not pass the mirror test were pigs (Marino & Colvin, 2015), apes and monkeys: gibbons, siamangs (Anderson & Gallup, 2015), rhesus macaques (Gallup, 1979). Dolphins (Reiss & Marino, 2001), elephants (Plotnik et al., 2006) and magpies (Prior et al., 2008) have also been claimed to pass the mirror test. However, these claims are controversial (Gallup & Anderson, 2018). Evidence for the dolphin and the elephant is based on a single individual, with evidence for the magpie being based on two individuals. None of the studies have been replicated, with magpies showing repetitive touching for long periods of time, something not seen in chimpanzees who habituated to the dyed body parts, therefore suggesting that other mechanisms may have played a role in the magpies' behaviours. Other criticisms involving mirror test results have ranged from attributing positive results to using conditioning techniques when researching primates, often using thousands of trials, using irritant colour dyes when testing fish, or employing the subjective scoring of video materials in relation to a dolphin (Anderson & Gallup, 2015; de Waal, 2019; Gallup & Anderson, 2018). However, Baragli et al. (2021) tested 11 horses for self-recognition in the mirror test, and demonstrated evidence for contingency behaviour (using the mirror as feedback for their own behaviour, such as sticking out their tongue), and passing the mark test (thus passing two out of five criteria listed by Mitchell (2012) as present in humans when passing the mirror test, with engaging in self-exploration, verbally identifying the image as oneself, and evoking self-conscious emotions such as embarrassment being the remaining three criteria).

It is uncontroversial that some species can use a mirror as a tool. For example, Marino and Colvin (2015) report that seven out of eight pigs were able to find food using a mirror, showing 'that the pigs do understand something about their own body as it is reflected in the mirror in relation to the hidden food' (p. 14). de Waal (2019) makes the point that mirror self-awareness may not be an all-or-none phenomenon but a continuum, with some species having some of the components for self-awareness as measured by the mirror test, and others fewer or none.

It is necessary to wield Occam's razor in self-awareness research to exclude anthropomorphic prejudice: only by avoiding subjective scoring and by replicating the mirror test under controlled conditions can we allow for an interpretation of self-awareness (e.g. see Cazzolla Gatti et al. (2021) in relation to the dog). Research on self-awareness also shows us that the precautionary principle may have to be applied: even if non-human animals only have some of the components for self-awareness, those components may make them aware of (aspects of) such emotions as fear, and our welfare decisions may have to take this into account.

2.7.4 Do Animals Recognise Internal States in Others, Like We Do?

We often assign intentions to animals, as in 'My dog gets the lead if it wants to go out for a walk'. That is only possible, however, if animals recognise internal mental states in others: 'My dog thinks that I think that if it gets the lead, I am inclined to go for a walk'.

What is the evidence for non-human animals recognising internal states in others?

Attributing mental states to others is part of a hierarchy of having access to internal mental states (Dennett, 1983; Dunbar, 2000):

0 order intentionality:	I cannot suppress or modify automatic reflection of my inner state (e.g. I suffer tissue damage and run).
1st order intentionality:	I believe X (e.g. that I am happy).
2nd order intentionality:	I believe that you believe X (this shows a reading of other minds, or a 'theory of mind' (e.g. Quesque & Rossetti, 2020).
3rd order intentionality:	I believe that you believe that I believe X (so I can use that knowledge to my advantage, and lie about it, or I can participate in pretend play).
4th order intentionality:	I believe that you believe that I believe that you believe X.
5th order intentionality:	(this is where most human participants start to get problems understanding what is going on; e.g. see Kinderman, Dunbar & Bentall, 1998).

...

The hierarchy of having access to internal mental states is important for many different reasons. Can animals, on purpose, lie, pretend to be something they are not (play dead, pretend to be bigger than they are), or does Occam's razor in combination with scientific evidence tell us that we are looking at automatic instinctual behaviour that has a pay-off? Where does the evidence point to?

According to Cheney and Seyfarth (1990, p. 207), chickens and vervet monkeys have first-order intentionality, but nothing beyond. For example, male vervet monkeys migrate to other groups. If a male vervet monkey migrates to another group that is habituated to humans, the new (non-habituated) group member quickly accepts that humans are not dangerous. It stops giving alarm calls and thus shows conformity to the group. If, however, a male vervet monkey that is habituated to humans migrates to a group that is not habituated to humans, the new (habituated) group member does not accept the alarm calls within the group but ignores them. In the latter case, the new group member does not conform to the group. So, what is the reason for this difference in conformity? According to Cheney and Seyfarth (1990), the new male in the latter group does not realise that there is a discrepancy between his knowledge and group knowledge: the male in the latter case has an internal state that is roughly equivalent to 'I know it's safe with humans around – I've seen the evidence before'. The male in the first case has an internal state that is roughly equivalent to 'I know it's safe with a human around – I've now seen the evidence'. So, in both cases it is possible to assume first-order intentionality: the vervet monkeys flag up what they believe to be the current state of affairs. There is no need to assume that a new human-habituated male in a group with non-habituated vervet monkeys thinks: 'They know x, but I know y, and I know that what they know is wrong, and that I am right'. This is Occam's razor shaving off second-order intentionality.

Anthropomorphic prejudice can be investigated at each level of intentionality. For example, Chapter 3 (this volume) will investigate evidence for first-order and second-order intentionality in animal communication and will consider anthropomorphic prejudice in these areas.

This section has shown that it is important to guard against anthropomorphic prejudice in relation to pain perception, emotion recognition, self-awareness, and to recognising internal states in others. Anthropomorphic prejudice may impact on our research and theory formation in these areas, and as a consequence, it may have an influence on human and animal welfare: for example, (inadvertently) compromising a dog's welfare may also impact on the welfare of a dog's carers (see Chapter 5, this volume). The four areas focused on in this section are not exhaustive. Other possible sources of anthropomorphic prejudice exist. For example, do animals have the same personality characteristics that we have (see Chapter 9)? Do animals communicate like we do (Chapter 3)? Are animals moral agents, as we are (e.g. Evans, 1906)? The areas of possible anthropomorphic prejudice are manifold.

2.8 Conclusion

As we have seen, our complex, ambiguous and ubiquitous relationship with other animals could be explained by several theories: the biophilia hypothesis, neoteny preference, the animal connection, social support theory, attachment theory and the biopsychosocial model. Given that the theories presented come from a wide variety of academic disciplines, an extensive critical evaluation of these theories is beyond the scope of this book. However, it is worth pointing out that none of these theories is without its critics (e.g. biophilia hypothesis criticism: Joye and De Block, (2011); biopsychosocial model criticism: Alvarez et al. (2012)). Nevertheless, all subsequent chapters in this volume provide a testing ground for these general theories on HAI.

The second part of this chapter considered anthropomorphism, explanations for its existence and, most importantly, consequences of anthropomorphic prejudice in studying HAI. After looking at possible sources for anthropomorphic prejudice (pain perception, emotion recognition, self-awareness and awareness of internal states in others), it was argued that systematic scientific research is imperative in HAI to test our theories and avoid anthropomorphic prejudice.

2.9 Learning Outcomes

Source: Illustrated by Loïc Maréchal adapted from istockphotos.

Chapter Summary

- The area of human–animal interaction studies the influence of non-human animals on us, as well as our influence on them.
- Our relationship with other animals is complex, ambiguous and ubiquitous. Theories in human–animal interaction should explain all of these elements.
- Anthropomorphism can stand in the way of studying human–animal interaction. We need objective empirical research to avoid and understand our prejudices in human–animal interaction.
- The inclination to anthropomorphise is ingrained in us, is subject to individual differences, and predicts our attitude to environmental protection, animal welfare and even to eating meat.

Check Your Understanding

Multiple Choice Questions

Question 1: How does neoteny explain our relationship with other animals?

- Answer a: we focus on animals that remind us of neoteny features in our own species and tend to ignore animals that do not have those features.
- Answer b: we focus on animals that we are attached to and tend to ignore animals that we are not attached to.
- Answer c: we focus on animals that we love (biophilia) and tend to ignore animals that we do not have a loving relationship with.

Question 2: How should we employ Occam's razor in HAI?

- Answer a: to be cautious in determining when animals have pain: to assume they have pain if there is a lack of evidence.
- Answer b: to prefer a simpler explanation over a more complex explanation when trying to explain animal behaviour.
- Answer c: to not use anthropomorphic explanations if we can avoid them.

Question 3: Why is it relevant to look at pain perception, emotions and self-awareness in non-human animals?

- Answer a: knowing more about these topics tells us why we anthropomorphise.
- Answer b: knowing more about these topics could lead to improvements in animal welfare.
- Answer c: knowing about these topics tells us why animals are like us.

Open Questions

This chapter makes a case for objective empirical research in the area of human–animal interaction.

1 What methods can you detect in this chapter on how to study human–animal interaction?
2 What principles are necessary to guide our research in this area?

3 Human–Animal Communication

Emile van der Zee and Kun Guo

3.1 Introduction

Human–animal communication is a core aspect of human–animal interaction (HAI): in any HAI information is exchanged between us and other animals. Communication not only involves speaking but also reading each other's behaviours, body posture, facial expressions and so forth. Any prejudices in human–animal communication, or our theories on this, may impact on animal welfare: if we comprehend other animals' intentions in the wrong way, their pain, their needs, their emotions and what disadvantages them, the consequences for their welfare – but also our own welfare – can be severe.

This chapter describes what communication is, how humans communicate with each other, how other animals communicate with conspecifics, and how communication features compare between humans and other animals. The last part of this chapter focuses on the overlap in the biological systems of humans and dogs, as a case study on how the biological systems of humans and dogs constrain communicative interaction.

Communication is the transfer of information from a signaller to a receiver. Information sent by a signaller must be meaningful or interpretable for the receiver. Communication channels vary widely: grunts, calls, songs, gestures, facial expressions, the use of colours, chemicals (e.g. pheromones), touch and so on. Intention or conscious processing is not necessary for communication to take place. For example, plants can signal the presence of predation to each other: damaged sugar maple seedlings and corn can emit airborne substances to signal damage, making it possible for other plants to become less palatable (van der Zee & Weary, 2010). Producing meaningful signals comes at a cost, however: there is an energy investment for the signaller, and signalling may make the signaller more vulnerable. Signallers have many ways to offset those costs: it may give them a reproductive advantage, it may give those genetically close to them an advantage in outcompeting others, or signallers may communicate messages that are not truthful and this therefore gives them a resource advantage. For example, subordinate tufted capuchin monkeys give more alarm calls signalling the presence of predators (felids, aerial predators and snakes) when dominant conspecifics are around, making it possible for these subordinates to have better access to a food source (Wheeler, 2009).

3.2 Information Constraints on Human–Animal Communication

The most efficient system that humans employ for communication is language, whether spoken or by using signs, for example, when using American Sign Language. Hockett (1960) lists 13 design features of what language is in order to compare humans and other animals, which he later expanded to 16:

DOI: 10.4324/9781003221753-3

1. Vocal-auditory channel: people use speaking/hearing.
2. Broadcast transmission and directional reception: a signaller broadcasts in all directions, but for a receiver the signal is associated with a particular direction.
3. Rapid fading: the signal only exists for a short time, and then fades away.
4. Interchangeability: signallers and receivers can take each other's roles.
5. Total feedback: signallers can monitor their output, and change it.
6. Specialisation: signalling is used for the purpose of communication, and not for other purposes (such as echolocation).
7. Semanticity: signals can be linked to specific meanings.
8. Arbitrariness: there is no direct connection between a signal and its meaning (the word *whale* does not look like a whale); the signal is a symbol.
9. Discreteness: language uses distinct units that can be combined using rules (e.g. the sounds /p/, /i/ and /n/ to form *pin* according to the rules of English phonology).
10. Displacement: it is possible to signal about things that are not present in space or time.
11. Productivity: it is possible to create new messages by combining known units of language.
12. Traditional transmission: language (learning) is influenced by culture.
13. Duality of patterning: meaningful signals (e.g. words) are composed of – and distinguished from each other by – meaningless units (e.g. individual sounds).
14. Prevarication: it is possible to make false statement (to lie).
15. Reflexivity: language can be used to talk about language (it can refer to itself).
16. Learnability: languages can be learned.

According to Hockett (1960), all human languages share these 16 features, but animal communication may have some, although not all, of them. For example, bee dancing may refer to the quantity and quality as well as the distance and direction to a food source (semanticity), but it is continuous (not discrete). Gibbons can make danger calls (discreteness and semanticity) but cannot coin new utterances (productivity). Hockett (1960) notes that these features are not always independent: for example, arbitrariness and duality of patterning are only possible if there is semanticity. He states that we share the first nine features with the great apes, but that the final seven features are uniquely human, due to our unique genetic make-up.

Hockett's (1960) system is a good starting point for looking at language in a comparative fashion, but some of his features need a more precise definition (e.g. how is productivity defined?), some features are missing (e.g. signing and touching as a way of communication), and other features have been introduced since his system was first published (e.g. intentionality; see Chapter 2, this volume). Most importantly, the concept 'language' needs to be defined more precisely.

The linguists Chomsky (1986, 2014) and Jackendoff (1993, 2002, 2010) provide formal definitions of language, for example, by making a distinction between phonology (sound structure), syntax (sentence structure), and semantics (meaning). Crucially, the rules for combining sounds are different from the rules for combining words into syntactic structures that form sentences, and these again are different from the rules composing meaning. In addition, rules can be identified that link together sound structures, sentence structures and meaning structures (e.g. Jackendoff (1993) for a basic introduction to all of these rule systems). From a linguistic viewpoint, words for humans, once learned, are most of the time long-term memory elements that link together phonological, syntactic and semantic information. For example, the sound structure /pin/ is linked to the syntactic structure 'noun' and the semantic structure $PIN_{token/type}$. This makes it possible, for example, to convert the thought of a particular pin (PIN_{token}), or pins in general (PIN_{type}), to a verbal expression by putting the /pin/ sound in a noun position in a sentence. Although words most of the time contain syntactic information for humans, this does not have to be the case. For example, *ouch* flags up pain, but *ouch* cannot be categorised syntactically. Sections 3.3 and 3.4 of this chapter will investigate how words and sentences are constructed by humans, what possible animal equivalents exist, and what role intentionality plays in communication. Intentionality is an important feature in human communication, since the main purpose of communication for humans is to change each other's internal states: for example, to inform, to persuade, to criticise or to flatter.

3.3 Knowledge of Words

All the words we know as individuals are stored in our mental lexicon. An average 17-year-old English speaker knows around 60,000 words (Bloom, 2001), but education level inflates this set, as well as speaking another language; in the latter case by at least around 10,000 words to be functional. At the age of two, word learning is exponential (Kauschke & Hofmeister, 2002), with children learning up to 12 words per month at the height of word learning, sometimes only needing one example of a mapping between a word and what the word refers to: so-called fast mapping. Children use theory of mind to learn words (see Chapter 2, this volume): for example, children attend to things in shared communication that the other party attends to ('joint attention'), including someone's pointing or gaze or even their emotional expressions (Bloom, 2001; Zheng et al., 2018). A minimum of second-order intentionality also plays a role in turn-taking, for example, when humans alternate with other speakers between questions and answers, or when we tell each other stories (Rossano, 2018).

Words belong to syntactic categories. Humans do not switch noun and verb positions in a sentence, nor put a noun in front of an -ed ending in English, which flags up past tense. The first example refers to syntax (how words combine to form sentences), and the second example refers to morphology (how meaningful word parts combine to form words). Morphology is not considered further in this chapter. Section 3.4 considers our knowledge of syntax in some detail. Every language has verbs and nouns, but not all languages have syntactic categories like prepositions or adjectives as we know them in English. In this section we will investigate whether or not animal calls or gestures are different from human words.

3.3.1 Animal Calls

Vervet monkeys can produce and comprehend at least three different alarm calls: calls for leopard, eagle and snake (Seyfarth et al., 1980). By hiding a loudspeaker that emitted recorded calls, Seyfarth, Cheney and Marler were able to determine that vervet monkeys have specific behavioural reactions to these calls: to stand on their hind legs and look around after a snake call, to flee into a tree after a leopard call, and to hide in the bushes after an eagle call. The context in which the calls took place (predator present versus absent), alarm call length, amplitude and a caller's age or sex had no impact on these responses. The authors concluded that their so-called play-back experiments showed a form of semantic signalling in vervet monkeys: a straightforward sound to meaning mapping. Syntactic categories did not play a role.

Vervet monkeys may use some alarm calls to also signal between-group and intra-group aggression (Price et al., 2015), with in some cases calls being related to signalling aggression only. However, the total repertoire of the vervet monkey is limited to fewer than ten, whereas the human mental lexicon contains around 60,000 words. Are there animals with larger vocabularies than vervet monkeys?

Savage-Rumbaugh et al. (1986) tried to teach a chimpanzee called Kanzi as many symbols as possible, using a keyboard. Kanzi was able to use up to 80 different symbols when almost four years old. Kanzi could indicate to want a banana, to be tickled and so forth. Kanzi appeared to be able to understand around 55 symbols and produced just over 45 symbols by pressing keyboard keys.

Other species have been shown to master larger artificial vocabularies than primates. Pepperberg (2002) has shown that a grey parrot called Alex could produce and understand more than 100 words, referring to objects, colours, shapes, object material and quantities up to eight. Pilley and Reid (2011) tried to push to its limits the number of words a Border collie called Chaser could understand. Over a period of three years, and up to four to five hours per day, Chaser was trained to pick up 1,038 objects upon being asked to do so by using unique names for these objects. Chaser demonstrated to understand 1,022 words and could also pick up objects belonging to a particular category: a toy, a ball or a frisbee.

These findings show that even under artificial conditions, the call repertoire of animals is more restricted than in humans. But is there also a qualitative difference?

Seyfarth and Cheney (1993) claimed that monkey calls are not intentional. They observed, for example, that vervet monkeys keep on calling even after behavioural responses have been given, thus ignoring the cognitive state of the respondent. Furthermore, in a set-up where mothers cannot see food or a predator, rhesus monkeys do not give more calls for 'food' (apple slice) or 'predator' (technician coming in to capture a monkey) if infants cannot see the food or the predator, compared to when they can see these things. This suggests that monkeys lack a theory of mind: they do not take into account the mental state of another monkey when calling. Monkeys appear to display only first-order intentionality in communication: they communicate what they believe to be the case.

Pepperberg, Gardiner and Luttrell (1999) suggested that also Alex, the grey parrot, did not have a theory of mind. Just like vervet monkeys, Alex kept on repeating his requests if he did not get what he wanted, and rarely accepted alternatives (which suggests that he used a fixed call-to-item correspondence). However, children modify their message if they do not get what they want: they may persist, but their requests take on a more flexible approach to convince the receiver to give them what they want.

The evidence so far suggests that apart from humans, only chimpanzees are aware of the mental states of others when making calls. Crockford, Wittig and Zuberbühler (2017) conducted an experiment in which chimpanzees either heard a rest hoo or an alert hoo from a loudspeaker hidden in the bushes, indicating the position of another chimpanzee. If after a few paces they saw a snake, and still heard a rest hoo coming from the bushes, they were more likely to call, monitor the speaker and mark the position of the snake by turning towards it, compared to when they heard an alert hoo coming out of the bushes. This shows second-order intentionality: chimpanzees took into account what another chimpanzee was assumed to know and adjusted their calls and other behaviours accordingly.

There are other qualitative differences between humans and animals. For humans, words can refer to things in the past, in the future, and to what we think about ('tomorrow', 'idea' and so forth). Animal calls, however, reflect more direct needs. For example, calls used by baboons (Hammerschmidt & Fischer, 2019), bonobos (Bermejo & Omedes, 1999), and monkeys mainly relate to food or predators, although larger call sets may include reference to such things as anger, status and maintaining proximity.

There also appears to be a qualitative difference in relation to word learning. As discussed before, children can learn words through fast mapping: learning the meaning of a word after only hearing one example. Fast mapping has been investigated in relation to the dog. Kaminsky, Call and Fisher (2004) argued that a Border collie called Rico showed evidence of learning the meaning of a word based on only one trial. Rico was asked to retrieve an object in an adjacent room from among seven known and one new object by giving him a new word. After having retrieved the new item based on the new word, Rico was tested again immediately afterwards and after four weeks. Rico correctly retrieved the new items using the new words when tested immediately after learning the new word-new object pairings but was at chance after four weeks. However, when making mistakes after four weeks, he did retrieve other non-familiar items and did not bring back the items he knew. If Rico was given a refresher recognition task after four weeks, he could bring back the correct new items upon hearing the new word. Based on knowing 200 words, and his fast-mapping performance, Kaminsky, Call and Fisher concluded that Rico had the word knowledge of a three-year-old child. The Rico study has been criticised, however, because among other things it was not necessary for Rico to focus on the features of the new object when learning a new word, but rather to focus on the features of the objects he knew, and therefore pair a new word with 'something else' using a process of exclusion (an interpretation that is strengthened by Rico confusing the meaning of a new word as referring to other unfamiliar objects after four weeks, apart from when trained to recognise the new object again) (Bloom, 2004; Kaulfuß & Mills, 2008; Markman & Abelev, 2004). Using appropriate control procedures to eliminate dogs learning words through exclusion, Fugazza et al. (2021) showed that dogs could only learn something close to fast mapping (four exposures to learn a new word) if taught in a social context and if they had a history of word learning, but that knowledge for the new words then deteriorated quickly (being tested again after ten minutes and after one hour). Current evidence thus suggests that fast mapping, whereby one example is shown in order to learn a new word, is a human trait.

Other areas of word learning in animals are less explored. Children and human adults have a shape bias when learning words referring to objects: when they know 50–100 words, humans start associating words referring to objects with the shape of an object, and not its texture or size (Landau et al., 1988). When understanding that a nonsense word like *dax* refers to an object with a particular shape, texture and size, and asked to pick out another dax, most children and adults select an object that resembles the original object in terms of its shape, and not its texture or size. Nonsense words were used in the dax experiment so as not to prejudice children and human adults with any known associations between words and objects. Dogs, however, do not seem to associate a word referring to an object with the shape of an object. van der Zee et al. (2012) asked a Border collie called Gable to pick up a dax or a gnark after having learnt the shape, size and/or texture of these objects, and showed that Gable tended to associate the meaning of these words with the size of an object. Although based on only one individual dog, this suggests that dogs may have a different word meaning in mind than humans do. More research on the shape bias in word learning in animals is necessary to determine whether qualitative differences between animals and humans exist.

Considering research on word learning in the dog, there appears to be another qualitative difference between human word learning, and word learning in dogs. Fugazza et al. (2021b) show that despite an intensive three-month training programme of word learning, only one dog out of 34 dogs that were naive to word learning but six out of six dogs that had substantial experience in word learning were able to learn new words (at least ten new words over the three-month period). The authors interpret this as evidence that only some dogs are exceptional word learners, in the same way that only some humans have absolute pitch hearing. Köszegi et al.

(2023) show that, although 20 dogs with a low vocabulary understanding of only five words each on average did not show evidence of individual word learning (with one exception), these dogs as a group taken together did show evidence of word learning as measured in a fetch task (with the overall effect not being attributable to the one exceptional dog). Köszegi et al. (2023) take this as evidence for implicit word knowledge in the low ability group, possibly comparable to a sensitivity to statistical regularities in children in the very early stages of word learning. This suggests that the vast majority of dogs, being of low ability in word knowledge, seem to be arrested in their word knowledge development at a level that is well below the three-year-old child level originally proposed by Kaminsky et al. (2004).

Even though there are quantitative and qualitative differences between human words and animal calls, the question presents itself whether despite these differences, humans are able to understand some naturally produced animal calls. Debracque et al. (2022) investigated whether humans may understand affective calls signalling affiliation (food-associated grunts), threat (aggressor barks), and distress (victim calls in social conflicts) in bonobos, chimpanzees and rhesus macaques. They explored whether there are elements in the acoustic patterns of all of these primates that make it possible for us to determine the meaning of these calls, or whether we might only be able to guess the meaning of these calls in primates closer to us on the phylogenetic scale (chimpanzees and bonobos as opposed to rhesus macaques). They found evidence for both hypotheses: apart from threat calls, human participants could categorise (A vs. B) and discriminate (A vs. non-A) the calls in all primates, although they were more accurate in relation to chimpanzee calls.

It could be argued that considering words, calls and symbols is too restrictive, and is a human-focused approach to communication. For example, chimpanzees also spontaneously use gestures in communication. Let us therefore look more closely at gestures in great apes.

3.3.2 Gesture Use in Great Apes

Byrne et al. (2017) have shown that there is an overlap in the gesture repertoires in great apes. Out of the 84 gestures that are produced without human intervention by all of the great apes together, 34 gestures overlap between the different species (e.g. waving arms, biting). Orangutans have a gesture repertoire of 29, five of which overlap with other great apes, gorillas have a repertoire of 37, ten of which overlap with other great apes, chimpanzees have a repertoire of 26, 15 of which overlap with other great apes, and bonobos have a repertoire of 21, 14 of which overlap with other great apes. Therefore, there is a possibility of finding a core set of gestures that are shared by humans and other great apes, like the chimpanzee. However, before spontaneous human and chimpanzee gestures can be compared, as Kersken et al. (2019) did, it is necessary to look first at the details of chimpanzee gestures.

Chimpanzee gestures can refer to such things as: *follow me*, which causes the signaller to be followed; *climb on me*, which causes an infant to climb onto an adult carer's body; *sexual attention*, which causes a male to respond sexually to a signaller; *stop that*, which causes cessation or a change in current behaviour (Hobaiter & Byrne, 2014). Byrne et al. (2017) argued that chimps wait to see the effect of a gesture ('response waiting'); if no response arrives, they persist – but otherwise they stop. Also, chimps who are shown a desirable food, persist and elaborate their gestural signalling if a keeper is reluctant to give all of it, but never persist and elaborate if they get what they wanted. Chimps are also more likely to use a silent visual gesture with an audience looking at them and are more likely to use a contact gesture (touching) when no one is attending; there is no difference in audible gestures (e.g. clapping) in such cases.

Observations like these led Byrne et al. (2017) to conclude that gesturing in chimpanzees is intentional: gesturing takes into account the internal state of a receiver.

In order to compare spontaneous human gestures and chimpanzee gestures Kersken et al. (2019) looked at the gestures in children who are two years old (who have not yet started to rely on verbal behaviour) in a German nursery, and in Ugandan children in their home environment, to rule out contextual and cultural effects. Chimpanzee data were based on 782 gestures of infant chimps one to five years old, and 2,129 gestures of juveniles five to ten years old. Human and chimpanzee gestures were only included if they showed intentionality: when they were targeted at a particular recipient. Kersken et al. (2019) found that 46 out of 52 gestures that were produced by children were also produced by chimpanzees: for example, raise an arm, shake an arm, grab and pull, hit an object or the ground with both hands, jump, push, and extend an arm with the palm vertically or upwards with all fingers open. Out of the six remaining gestures where there was no overlap with chimpanzees, four gestures were also found in the gorilla repertoire. Although there was a high (89%) overlap between the gestures in young children and chimpanzees, there was a qualitative difference in the gestures produced. Children produced more referential (pointing) gestures, and whereas children produced more silent visual gestures (raise arm, wave hand), chimpanzees produced more audible gestures (hit the ground, clap). The authors concluded that there may be a universal gesture repertoire underlying human and chimpanzee gesturing, and qualitative differences may either indicate genuine species differences or be due to different environmental pressures. Chimpanzees are likely to be unique in the animal kingdom, however, in having an overlapping gesture repertoire with humans, and showing second-order intentionality in gesture communication.

In order to further test whether there is a universal gesture repertoire underlying human and chimpanzee gesturing, Graham and Hobaiter (2023) asked 5,656 participants to determine the meaning of chimpanzee and bonobo gestures in 20 different videos. Participants were either provided with a written context in which the gestures were made, or not. Gestures either had a unique meaning (e.g. Big Loud Scratch: groom me), or more than one meaning (e.g. Object Shake: let's have sex, groom me, move away). In a multiple-choice test to determine the meaning of the gestures in the videos, participants were accurate for nine out of ten gestures (only Object Shake was at chance level), even without contextual information. The role of context was not significant, but played a marginal role when gestures were ambiguous. Graham and Hobaiter (2023) interpret these findings as evidence that 'an ancestral system of gesture appears to have been retained after our divergence from other apes' (p. 1).

Do animal calls or gestures have syntactic categories attached to them? It would not be possible to categorise a vervet monkey's call for eagle unambiguously as a noun; it could refer to *hide* (a verb), *hide in the bushes* (a sentence), or *scary bird* (a noun phrase), and so forth. Whether spontaneous gestures in the great apes, like *follow me*, can be considered verb–noun combinations requires a more sophisticated look into what syntax is. Section 3.4 focuses on this.

3.3.3 Conclusion

Human knowledge of words seems to be quantitatively and qualitatively different from knowledge of calls and gestures in animals: call and gesture repertoires are smaller in animals, expressions beyond the here and now seem absent, only chimpanzees seem to use second-order intentionality in call and gesture use, and word learning through fast mapping only seems to be a human trait. However, humans are able to determine the affective nature of several primate calls, as well as the meaning of many gestures in some great apes. More research in these areas is necessary.

3.4 Knowledge of Syntax

Already in the seventeenth century Descartes argued that humans are creative language users, whereas animals can at best mechanically repeat what human beings say, like parrots do (Descartes, 1637, part IV). Hockett (1960) and Chomsky (1986) generally follow Descartes' idea that the way in which humans produce language sets us apart from other animals. However, Chomsky is quite specific in what he means by knowledge of syntax (Chomsky, 1986; Fitch et al., 2005; Hauser et al., 2002). This section will consider this idea.

The total number of meanings that can be expressed by a sentence is infinite: theoretically, human beings can think and talk about anything for any length of time. This means that syntax should allow us to make any possible sentence, of any possible length; of course, within the rules that we deem acceptable for a particular language. So, what are the formal properties of syntax according to Chomsky (1986, 2014) that make this possible?

The easiest way to think of syntax is that it allows us to produce patterns inside of patterns (Jackendoff, 1993). This is also known as recursion. If 'S' is a sentence, and one were to say *Max likes that S*, then we have a sentence inside a sentence, for example, [Max likes that [Ben plays football]$_{S1}$]$_{S2}$. This pattern, however, can be put into the same pattern: [Zoë thinks that [[Max likes that [Ben plays football]$_{S1}$]$_{S2}$]$_{S3}$, and so forth. Recursion therefore makes it possible to generate sentences of infinite length in principle. It can be argued that our short-term memory may lose track of what is being said, but technically it is possible to go on forever producing a sentence.

Recursion is not only possible at sentence level but also at phrase level. This is explained best by putting the sentence template *Max likes that S* into a tree diagram (Figure 3.1).

In Figure 3.1 *Max* is a noun which forms a noun phrase, and *likes* is a verb forming a verb phrase. Phrases can be expanded forever, in a similar way as sentences can: [Max]$_{NP1}$ [the head of state]$_{NP2}$ [top of his class]$_{NP3}$ [owning a blue garden shed]$_{NP4}$ likes that S]$_{S}$. In NP_2 *state* is a noun forming a noun phrase (NP), in NP_3 *class* is a noun forming an NP, and in NP_4, *shed* is a noun forming an NP. And of course, an infinite number of NPs could have been added. Also, prepositional phrases can be expanded forever: [George is [in the garden] $_{PP1}$ [with his

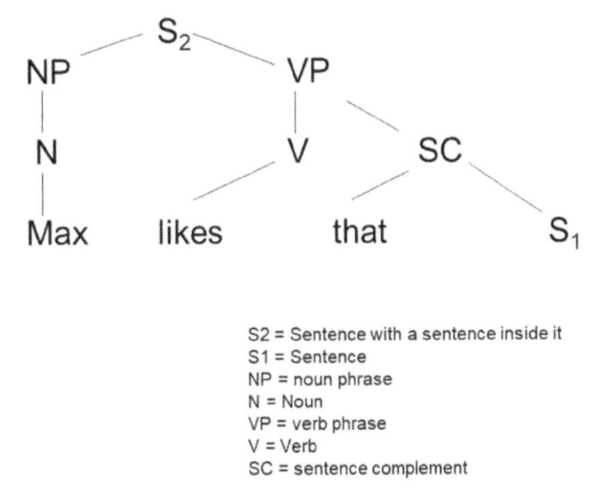

Figure 3.1 A tree diagram of the sentence template 'Max likes that S'.

wellies]$_{PP2}$ [under a parasol]$_{PP3}$ [next to Marc]$_{PP4}$]$_S$. In PP_1 *in* is the preposition forming a prepositional phrase (PP), in PP_2 *with* is a preposition forming a PP, and so forth. What is also important, however, is that the structure in Figure 3.1 appears hierarchical – some elements are higher up, and other elements are lower down. Apart from recursion, hierarchy is also an important feature of syntactic structures.

The reason why hierarchy is important in syntax can be illustrated by looking at the following ambiguous sentence: *Katie eats the burger with a skewer*. This sentence can mean that Katie eats a burger with a skewer embedded in it, and it can refer to Katie eating a burger by using a skewer (In fact, the sentence is even more ambiguous. It can also refer to Katie eating a burger while being accompanied by a skewer, as if the skewer were a person, and it can refer to Katie eating a burger and a skewer eating the same burger. We will not look at these latter two readings here, which are also partly based on syntactic structure). Both these meanings are represented by the same sentence (surface structure), but deeper down the syntactic structures are different (deep structure): see Figures 3.2 and 3.3.

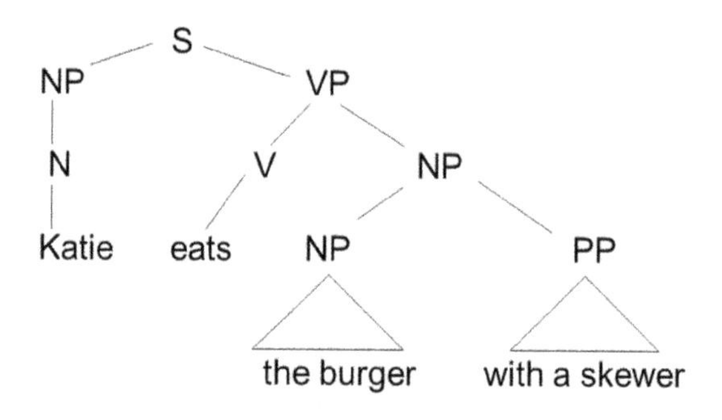

Figure 3.2 The syntactic structure representing the meaning that Katie is eating a burger which has a skewer in it. Triangles reflect not looking into further syntactic detail.

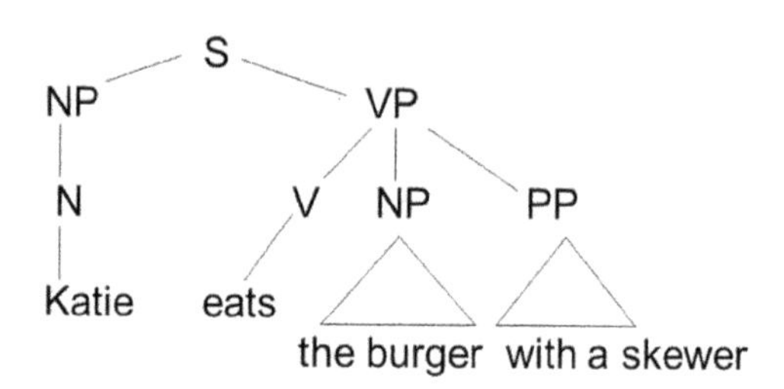

Figure 3.3 The syntactic structure representing the meaning that Katie is eating a burger by using a skewer.

In Figure 3.2, *with a skewer* is a property of *the burger*, syntactically represented by the fact that both phrases are part of an NP, are sisters (they are at the same level in the hierarchy), and that *with a skewer* is to the right of the NP it modifies (English is right branching; languages can also be left branching). In Figure 3.3, both *the burger* and *with a skewer* are properties of the verb *eat*; they are both sisters and to the right of the verb under the VP node. These diagrams thus show that hierarchical structure is an important part of syntax: hierarchical structure helps build meaning. The meaning of words in a sentence, together with syntactic deep structure deliver sentence meaning. This is referred to as compositionality. The formal question is thus, do other animals also show signs of recursiveness, hierarchical structure and compositionality?

3.4.1 Recursiveness, Hierarchical Structure and Compositionality in Animal Calls?

In order to investigate the properties of recursiveness, hierarchical structure and compositionality in animals, we first consider animals that can produce call sequences, starting with songbirds, which produce long call sequences (Suzuki et al., 2019). Bird vocalisations evolved at least three times independently: in songbirds, parrots and hummingbirds, together covering more than half of the >10,000 bird species (Clayton et al., 2009). Hallmarks of songs from songbirds are: learnability of the songs through exposure (so not preprogrammed innate sound sequences), plasticity of brain regions that support song (e.g. growth of brain areas involved in singing when doing this more frequently), and the use of a syrinx (a bipartite voice box) to produce complex sounds (some species can even produce two different sounds at the same time). Two studies looking at recursiveness, hierarchical structure and compositionality in songbirds are considered next.

Suzuki et al. (2016, 2018) observed that the Japanese tit can use over ten different notes in their vocal repertoire which can be produced singly or in combination. ABC call sequences are produced by signallers to warn conspecifics about predators – receivers scan for danger. D calls attract conspecifics – receivers approach. ABC-D sequences are used to recruit conspecifics to mob a stationary predator – receivers scan and approach to the same extent as for ABC and D alone. In an experiment, Suzuki et al. (2016, 2018) artificially introduced D-ABC sequences and observed that it resulted in reduced scanning and barely any approaching behaviour. Since Japanese tits behave differently for ABC-D call sequences compared to D-ABC sequences, Suzuki et al. (2016, 2018) drew the conclusion that because of ordering differences, these songbirds assign a different meaning to the songs, and that they therefore show signs of compositionality. Suzuki et al. (2018) did state, however, that there is no evidence of a hierarchical structure in the songs, nor evidence of free productivity (recursion).

Engesser, Ridley and Townsend (2016) also claimed evidence for compositionality in the southern pied babbler. This songbird does not sing, but produces 17 discrete calls, among which are noisy alert calls in relation to low-urgency threats (e.g. hares and antelopes), and repetitive recruitment calls resulting in receivers approaching, overall group movement, counter calling, or no response. Both call types can be combined into a mobbing sequence to signal the presence of mostly terrestrial predators (mongooses, snakes or foxes). The mobbing sequence consists of one or two alert calls (two if the alert calls are short), and four to seven recruitment calls. Vigilance and latency to resume normal behaviour are almost absent for individual alert calls, slightly increased for recruitment calls (with attention to the caller) but very much increased for mobbing sequences. A sequence where a chuck call (used when foraging) artificially replaced the alert call in the mobbing sequence did not lead to an increase in vigilance or latency to resume normal behaviour. The authors interpreted their results as evidence for rudimentary composition: vigilance times and latency times to resume normal behaviour in mobbing sequences are not simply an addition of alert calls and recruitment calls but show a much higher increase. They also imply that hierarchical structuring is not present.

Both songbird studies can be interpreted as weak forms of compositionality: sequential ordering (Figure 3.4). Sequential ordering is often used in the comparative communication literature to indicate some level of signal organisation. As has been shown in the foregoing, however, sequential ordering is not how human syntax works: a sequential order describes surface structure, but not deep structure – and also deep structure can have consequences for sentence meaning (compare Figure 3.4 with Figures 3.2 and 3.3).

Zuberbühler (2019, 2020, 2021) argued that there are no natural signalling systems in other animals that go beyond a combination of two signals as illustrated in Figure 3.4. As he showed, at best combinatorial signalling in even primates is made up of two signals forming a composite signal. Zuberbühler (2020) reported that bonobos produce five types of calls related to food: bark (B), peep (P), peep-yelp (PY), yelp (Y), and grunt (G). These calls can be combined: call pairs with B or P were linked to kiwi (high food quality), whereas call pairs with Y or PY were linked to apple (low food quality). Since such call pairs are hidden inside larger sequences, it is not easy to detect such pairs, but other calls when emitted seemed irrelevant. Zuberbühler (2020) speculated that call pairs serve to disambiguate the message: single calls can sometimes be used for non-food items, and the number of food items that bonobos encounter in the wild is potentially large, so that single calls work less well than call pairs to be more specific about which food items are encountered.

Ramos and Ades (2012) confirmed that the dog also understands two-signal combinations well: a female mongrel called Sofia, was able to carry out simple object-action requests (*ball fetch, stick point*) flexibly across contexts. Pilley and Hinzmann (2014) claimed that Chaser could understand three signal combinations, for example *to ball take Frisby* (as opposed to *to Frisby take ball*, which would have resulted in a different action sequence). However, at least two observations suggest that Chaser carried out two command sequences in a row and was not reacting to a three-signal combination. First, there was a one- to two-second break between sequences like *to ball* and *take Frisby*, suggesting that Chaser did not necessarily need

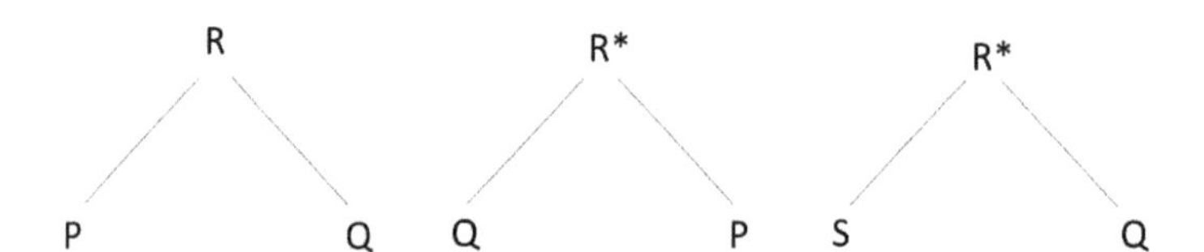

Figure 3.4 Sequential ordering or weak compositionality where two signals are combined into one signal, and where reversing the order or signal substitution leads to unacceptable signal strings. In the first example a composition of P (e.g. the ABC combination of calls in the Japanese tit, or the alert call in the southern pied babbler) and Q (e.g. the D call in the Japanese tit, or the recruitment call in the southern pied babbler) leads to an acceptable combination R. In the second example, switching around the signals for a combination sequence (e.g. introducing the D-ABC sequence for the Japanese tit) leads to an unacceptable signal string R*. And in the third example, R* is made up of existing signals S and Q, but is unacceptable since S is not normally combined with Q (e.g. a chuck call followed by recruitment calls in the southern pied babbler), since P is normally combined with Q.

to understand a prepositional object-verb-direct object sequence as one unit, as argued, but listened to *to ball* and *take Frisby* as separate units. Secondly, however, Pilley and Hinzmann (2014) report having tried to teach Chaser three signal sequences like 'ball, take, Frisby', using many trials, but never having succeeded. It seems that Chaser was not able to segment these three signal combinations into two separate commands. Based on current evidence, it therefore seems reasonable to assume that there is a two-signal boundary in the understanding of verbal information by the dog, congruent with Zuberbühler's hypothesis.

The literature on syntactic knowledge in animals reports many so-called artificial grammar experiments. The general idea is that birds, primates or other animals are taught an artificial grammar that is associated with an aspect of their behaviour, and that a behavioural change is observed when the rules of the grammar are violated. For example, zebra finches can be taught to peck a grey key when hearing an 'abcdefabefcdab' order of natural zebra finch syllables, but are taught to refrain from pecking when hearing 'abdfcedfabceab' (Chen & ten Cate, 2015). By changing syllables in both strings in these so-called Go-noGo experiments (in this context: peck or do not peck), it then becomes possible to see whether pecking increases or decreases, and thus whether the zebra finches are sensitive, for example, to changes in the locations of syllables, and the introduction of new syllables. By using an experiment of this nature, Abe and Watanabe (2011) reported that the Bengalese finch can process hierarchical structures. However, such conclusions have been contested. For example, Beckers, Bolhuis, Okanoya and Berwick (2012) argue that there was more acoustic overlap between familiarisation strings and grammatical strings, than between familiarisation strings and non-grammatical strings, making it easier for the finches to select the grammatical strings. Beckers et al. (2017) present an overview of the many acoustical confounds in artificial grammar experiments with birds. At present, there therefore does not seem to be any evidence for knowledge of hierarchical syntactic structure in the natural signalling repertoires of animals, or in animals involved in artificial grammar experiments.

3.4.2 Recursiveness, Hierarchical Structure and Compositionality in Animal Gestures?

Teaching chimpanzees American Sign Language (ASL) resulted from trying to teach chimpanzees English (Hayes, 1951), but discovering that their voice box was not suitable for producing the right sounds (Zuberbühler, 2020). Gardner and Gardner (1975) were the first to report an experiment in which a chimpanzee called Washoe was taught ASL. Washoe was able to use around 130 signs, but claims for creativity in language use were limited to very few anecdotal examples, such as signing *water* and *bird* upon seeing a swan, after Washoe was asked 'What that?'

To investigate the possibility of recursiveness, hierarchical structure or compositionality, Terrace et al. (1979) trained a chimpanzee called Nim Chimpsky (named after Noam Chomsky) in ASL from when he was two weeks old for a period of four years. If independent observers registered unprompted, non-moulded (sometimes trainers moulded Nim's hands to teach him the right sign), non-modelled (not a repetition of what a trainer had just signed before) signs for five days, they considered this a sign. With that criterion, Nim had the ability to produce 125 signs. Regularity in signing was more frequent in two sign combinations (e.g. more signing of *give x* than *x give*, *more x* than *x more*) than in three, four, five or more sign combinations. Young children tend to add information when going from the two-word stage (e.g. *sit chair*) to the three word-stage (e.g. *sit daddy chair*); repetition is rare. Nim's three or more sign combinations, however, contained redundant information. For example, three-word combinations were: *eat*

Nim eat, *play me Nim*, and *nut Nim nut*, with the most frequent four-word combination being *eat drink eat drink*. The longest sign sequence (16 signs) was: *Give orange me give eat orange me eat orange give me eat orange give me you*. The authors interpreted redundancy as Nim wanting to emphasise what he wanted. They furthermore observed some other interesting differences between sentence learning in children and in Nim: when mean-length utterances increase in children, complexity also increases, but not for Nim. Teacher imitations and partial repetitions in children go down during development, but not for Nim. Interruptions of 'teachers' signing go down in children during development, but not for Nim. In addition, the authors could not find evidence for hierarchical representations. Composition also seemed to adhere to the two-signal boundary intimated by Zuberbühler (2019, 2020, 2021).

When investigating call or gesture sequences, one has to be careful: signals can appear to be combined but are not, since they may be idiomatic (Townsend et al., 2018). A typical English idiom would be *once in a blue moon*, which refers to *rarely*. Idioms are special in that the meaning of the idiomatic expression cannot be derived from the meaning of the signal parts, plus the syntactic structure putting the signals together. Arnold and Zuberbühler (2012) argued that pyow-hack sequences in putty-nosed monkeys (consisting of two to eight calls) are idiomatic. Sequences are given by males prior to group movement, and sequence length has an impact on the receiver, but not on the relative contributions of the individual calls. Recursiveness, hierarchical structure and compositionality obviously do not play a role in idiomatic expressions.

3.4.3 Conclusion

There is no evidence for recursion or hierarchical structure in animal communication, unlike in human language, and there is only evidence for weak compositionality – combining two signals to form a third signal (unlike in human communication, compositionality in animals is not based on deep structure).

We have so far mostly focused on the features of animal communication, and human communication, although we have seen that there is some overlap in spontaneous human and chimpanzee gestures, that humans can understand some primate calls and gestures without training, that both humans and other animals can produce two-signal combinations with a meaning that goes beyond the meaning of the two signals in isolation, and that not only humans but also animals can use idiomatic (fixed meaning) signal strings. In order to fully understand communication between other animals and us, however, we also need to look at HAIs that are not based on calls and gestures. In the remaining part of this chapter, we do this by focusing on human–dog interaction. It will be demonstrated that human–dog communication is constrained by the biological make-up of these two species.

3.5 Biological Constraints on Communication

Humans make great use of visual and auditory information in communication and social interaction. Vision, in particular, is regarded as the most important human sensory modality because of its dominance in mediating the amount of perceived sensory information from our surroundings and proportion (> 50%) of neocortex involved in visual processing (Gazzaniga et al., 2014). From a human's perspective, we may expect other animals to also use visual and auditory cues when interacting with us. However, due to differences in the visual, auditory and brain structures humans and other animals are likely to show different sensitivities to the same visual and/or auditory inputs. In other words, animals may not see what humans see, and hear what humans hear or vice versa. Considering that interpreting the same information differently may lead to different actions (e.g. the same facial expression perceived as happiness or anger may induce approach or avoidance behaviour), for the purpose of safe HAI and animal welfare, it is crucial to understand our biological differences. Below, the biological constraints on communication are illustrated using human–dog interaction as a case study.

Due to their unique physical eye and ear structures as running predators, dogs tend to typically show better visual motion detection and hearing performance, but poorer spatial and colour vision capacity than humans (Barber et al., 2020a). These differences in sensory perception, in general, do not affect reliable and efficient social communication and interaction between dogs and humans. In fact, due to their unique history of domestication and selective breeding, dogs are unusually skilled at reading human social-communicative signals (Hare & Tomasello, 2005). From this perspective, the history of human–dog interaction may represent a case of convergent cognitive evolution (i.e. for two species with distant phylogenetic relations to have independently evolved similar cognitive skills as a result of their adaptation to the same environment). Indeed, recent studies have revealed both remarkable similarities between dogs and humans, and species-specific differences in social cognition. Using face, identity and emotion perception, three crucial components of social cognition, as examples, we will discuss how and to what extent dogs process visual and auditory cues to facilitate dog–human interactions. It is worth mentioning that we currently know very little about the neural mechanisms underlying social cognition in dogs due to various technical and ethical challenges. The experimental evidence discussed below is mainly from behavioural and eye-tracking studies. It is also worth stressing that breed diversification in dogs is associated with high variation in morphology (e.g. body height and size, nose length, eye location, ear shape) and ecological function (e.g. scent hounds vs. gun dogs), potentially leading to large breed differences in visual and auditory functioning (e.g. differences in visual field size and visual acuity between short- and long-nosed dogs; Barber et al., 2020a, 2020b). Hence caution is needed when applying a general evaluation of human–dog interaction to individual dog breeds.

3.5.1 Face Perception

Faces are probably the most important visual stimuli in human social interactions, as they can transmit multifaceted visual information about an individual's gender, age, familiarity, attractiveness, affective state, intention, cognitive activity (e.g. concentration), and temperament (e.g. hostility). Consequently, a face-specific cognitive and neural mechanism has been evolved/developed to process these complex facial cues in humans (Bruce & Young, 2012). Probably being aware of the crucial role of human faces in human–dog interactions, dogs are also tuned to human faces, exhibit high perceptual sensitivities to some facial cues (e.g. gaze direction), and demonstrate some human-like cognitive strategies in viewing faces.

Probably due to the right hemisphere advantage in face processing, humans often show left gaze bias (LGB) when viewing other peoples' faces, in which the left hemiface (from the viewer's perspective) is often inspected first and/or for longer time periods (Guo et al., 2012; Figure 3.5). Considering that six-month-old infants show a general, inherent LGB for both face and non-face object images, which later transforms itself into a specific LGB for face images only as demonstrated in 4-year-olds and adults (Guo et al., 2009; Racca et al., 2012), the

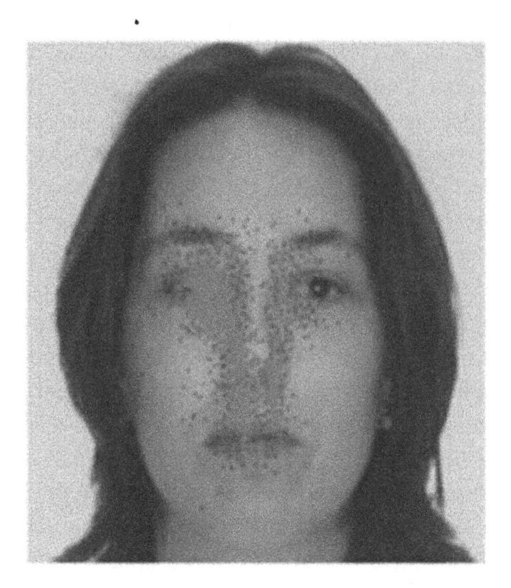

Figure 3.5 Demonstration of left gaze bias in humans when viewing other human faces. The figure shows overall distribution of the first fixation on a typical human face made by human viewers. Each red dot represents the first saccade destination sampled from one face-viewing trial.

Source: Adapted from Guo et al. (2012).

face-related LGB in humans is likely an acquired behaviour, possibly through the process of an experience-dependent gradual specialisation during development.

When viewing neutral faces of different species, dogs demonstrated both initial and overall LGB for human faces, but not for monkey or dog faces, nor for inanimate object images (Guo et al., 2009). Such face-specific and species-sensitive gaze biases in dogs may have significant adaptive value and could be linked to the dog's unique evolutionary and ontogenetic history, as the ability to extract information from human faces and responding appropriately could have had a selective advantage during the process of domestication. The lack of LGB in viewing monkey and dog faces might reflect a reduced need or sensitivity in assessing these faces with neutral expressions. Unlike humans, face-viewing gaze bias in dogs is further modulated by the viewed facial expressions. Dogs tended to show a consistent LGB for negative and neutral human facial expressions, but no bias for positive expressions. Perhaps dogs interpret human neutral expressions as potentially negative, given their lack of clear approach signals. They, however, demonstrated a differential gaze asymmetry for dog faces based on their emotional valence, with no gaze bias for neutral expressions but an LGB for negative expressions and a right gaze bias for positive expressions (Racca et al., 2012). These observations are broadly consistent with the valence model of cerebral lateralisation in emotion perception, with the left and right hemisphere mainly involved in the processing of positive and negative emotions, respectively (Demaree et al., 2005). Hence in comparison with humans, gaze asymmetry in dogs is a reflection of brain lateralisation in both face and emotion perception.

Another stereotypical face-viewing gaze behaviour in humans is a strong preference towards the eyes (eye bias). Among various internal (e.g. eyes, nose, mouth) and external facial features (e.g. hair style), humans tend to direct the majority of fixations or viewing time at the eyes

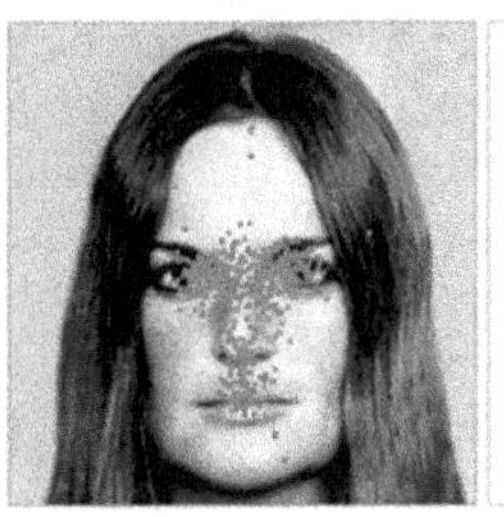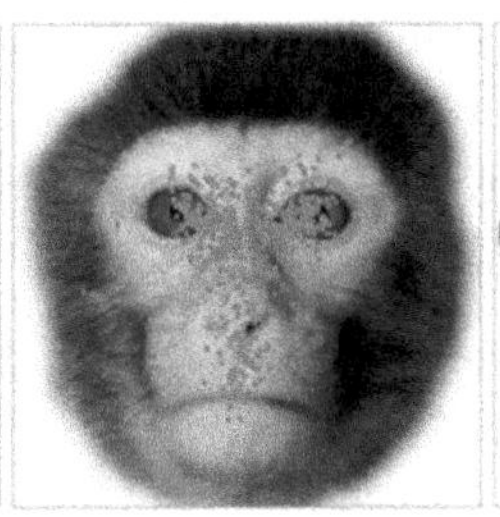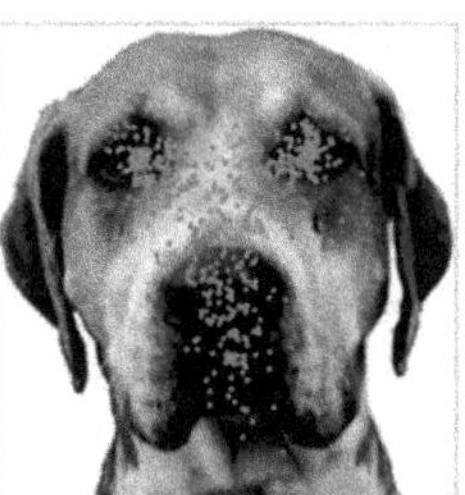

Figure 3.6 Demonstration of eye bias in humans when viewing human, chimpanzee, monkey and dog faces. The figure shows overall face-viewing fixation distribution on these faces. The red dots within each face image indicate the position of each fixation sampled during this face-viewing trial across all human viewers.

Source: Adapted from Guo et al. (2019).

when viewing faces irrespective of task demands (e.g. free viewing, face learning, identity judgement, facial expression categorisation) and face species (e.g. human, monkey, dog and cat faces), suggesting a crucial role of the eyes in transmitting various types of information and a generic 'built-in' scanning strategy in our brains (Guo et al., 2010; Guo, 2012; Figure 3.6). Like humans, dogs also prefer to look more at the eyes when viewing human and dog faces (Barber et al., 2016; Somppi et al., 2014). Even at early stages of development, dogs show a spontaneous tendency to gaze at human faces and to make eye contact in a wide range of contexts, such as seeking human help to solve difficult tasks and evaluating human gaze direction to locate hidden food or to adapt their behaviour accordingly (Hare & Tomasello, 2005; Siniscalchi et al., 2018).

Regarding the neural mechanisms underlying face perception, brain imaging studies have revealed three key face-selective areas in the human cortex, including the occipital face area, the fusiform face area and the posterior superior temporal sulcus face area (Bruce & Young, 2012). Even with various technical challenges, a few fMRI studies on dogs have reported enhanced brain activity in a localised region within the dog temporal cortex when viewing human or dog faces (Cuaya et al., 2016; Dilks et al., 2015). This localised 'face-sensitive' area in dogs is likely equivalent to the fusiform face area in humans, suggesting a high degree of evolutionary conservation of this brain region for face processing across animal species.

3.5.2 Identity Perception

Human faces provide a dominant role in identity recognition, and we are highly efficient in discriminating and identifying individuals' faces through a specialised holistic cognitive process, in which we perceive relations among individual facial features and then integrate all features into a single representation of the whole face. In other words, we recognise a face by seeing it as a Gestalt image, not as individual features, such as the eyes, nose and mouth. This holistic process is orientation-sensitive as typically demonstrated by the face-inversion effect, defined as a larger decrease in recognition performance for faces that are inverted rather than upright, whereas this is not the case for objects (Bruce & Young, 2012).

Dogs can use visual cues alone to discriminate face identities, possibly through a human-like inversion-sensitive holistic process. When two human or dog faces were simultaneously presented side by side in a preferential looking procedure, dogs looked longer at the novel human face than the prior-exposed familiar human face, but they looked longer at a familiar dog than a novel dog face, suggesting dogs may use species-specific facial cues to differentiate individual dogs and humans. Such looking preferences, however, were diminished when the faces were presented upside-down, suggesting dogs may also use a holistic strategy in identity perception (Racca et al., 2010). Dogs could also discriminate their owner from another familiar human in a real-life experimental setting (e.g. heads poking through a box) or in face pictures, but recognition accuracy was decreased sharply when a balaclava mask was applied, leaving only the eyes, nose and mouth visible, further indicating the importance of global facial cues or configural processes in the dog's face identity perception (Huber et al., 2013). Recent dog eye-tracking studies also revealed human-like differences in gaze distribution between viewing familiar and unfamiliar faces, with a greater number of fixations allocated at the eye region in familiar faces (Somppi et al., 2014).

Like humans, dogs can use auditory cues alone to discriminate individual identities, such as matching one's voice with a face (Siniscalchi et al., 2018). In a habituation–dishabituation paradigm, dogs showed the ability to spontaneously differentiate the same spoken phonemes across different unfamiliar speakers, suggesting dogs may be able to recognise individual people's voices (Root-Gutteridge et al., 2019). Playback experiments further indicated dogs can detect not only the semantic content of conspecific barks but also remember individual characteristics, such as body size (Faragó et al., 2010).

3.5.3 Emotion Perception

Understanding each other's emotional state has a crucial adaptive value in social interaction (e.g. a happy/angry emotion tends to prompt approach/avoidance behaviour). Because facial expression is the dominant channel of emotional expression in humans compared to bodily and auditory cues, we universally show a very high perceptual sensitivity to other people's facial expressions, especially those expressions that represent our typical emotional states (i.e. happiness, sadness, fear, anger, disgust and surprise). We can categorically discriminate these universal expressions even with very brief face presentation times (< 100ms) or when focal attention is not fully available (Bruce & Young, 2012).

Dogs are also visually sensitive to human facial expressions (Figure 3.7). They can visually discriminate human smiling from neutral faces (Nagasawa et al., 2011), happy from disgusted faces (Buttelmann & Tomasello, 2013), and happy from angry faces with above chance-level performance (Albuquerque et al., 2016; Müller et al., 2015). In a touchscreen-trained choice paradigm, dogs could selectively match top-halves and bottom-halves of unfamiliar human faces displaying happy or angry expressions, suggesting a human-like holistic process in perceiving human facial expressions by dogs (Müller et al., 2015).

Recent empirical evidence further suggests that dogs seem to understand the meaning of these discriminable expressions and change their actions or behaviour accordingly (Hare & Tomasello, 2005). In fact, when viewing faces of negative expressions, they often display stressful behaviour (e.g. mouth-licking; Albuquerque et al., 2018) and physiological responses (e.g. cardiac responses; Barber et al., 2017). When responding to humans who communicate

Figure 3.7 Video clip of the clever dog lab: Touchscreen training – discrimination of angry and happy faces (www.youtube.com/watch?v=Um30o-RT-YE).

directional cues, such as pointing, head turning, glancing or nodding, they often change their approach preference depending upon whether the perceived human expression is positive or negative (Ford et al., 2019).

Although dogs might perceive human emotions to a certain degree, there are differences in the cognitive strategies underlying visual emotion perception between the two species. Dogs often show non-human-like facial expressions to humans when reacting to emotionally comparable contexts (Caeiro et al., 2017). The little commonality in facial expressions between the two species may contribute to differences in their gaze distributions when inspecting the same facial expressions. Whereas humans allocated gaze according to the targeted facial features (e.g. eye bias), the viewed expression category and face species, dogs only modulated their gaze according to the targeted facial features (Correia-Caeiro et al., 2020). This lack of gaze sensitivity to the displayed expression category in dogs raises a question about their capability to discriminate different human emotion categories. On the other hand, human gaze allocation was not correlated with the diagnostic facial movements occurring in (especially dog) emotional expressions, which may be the cause of their poor performance in recognising dog emotions (Correia-Caeiro et al., 2020). Furthermore, when visual stimuli were changed from faces to whole figures (including both facial and bodily expressive cues), dogs attended more to the body than the head of human and dog figures, whereas humans focused more on the head of both species (Correia-Caeiro et al., 2021). Clearly, these two species use different gaze strategies to extract diagnostic expressive cues in either faces or full bodies for visual emotion perception.

Regarding auditory emotion perception, dogs, like humans, can use auditory cues to recognise emotions (Siniscalchi et al., 2018). They could spontaneously match unfamiliar happy or angry human and dog vocalisation with unfamiliar happy or angry human and dog faces, suggesting the existence of mental prototypes for (at least) positive versus negative emotion categories in their minds (Albuquerque et al., 2016). They could also discriminate emotionally relevant information and emotional valence (e.g. happiness versus sadness versus fear) of human communication based on auditory cues (Huber et al., 2017; Siniscalchi et al., 2018).

3.5.4 Conclusion

Dogs can use visual/auditory cues and visual-auditory integration to perceive (at least) human identity and emotion, with human-like underlying cognitive (e.g. holistic face processing, face-related cognitive bias) and neural mechanisms (e.g. face-selective cortical regions).

Although social cognition in dogs and humans may be qualitatively similar, there are marked quantitative differences between them. For instance, the constraints in the dog's visual system are likely to impose higher perceptual thresholds to perceive human identity and emotions (e.g. they may not be able to detect subtle human emotions); the species-specific communication and emotional expression processes may restrict the number of human emotion categories recognised by dogs or even prompt a valence- rather than category-based emotion perception in dogs. Humans need to take these differences into consideration when interacting with dogs; be aware that dogs may use a range of visual, tactile, acoustic and olfactory signals to communicate with us; and have a realistic expectation of dogs' ability to detect and discriminate our social cues.

Within these species-dependent social cognition processes, it is unlikely dogs or humans can use the same strategies to perceive the same social cues from both species and achieve comparable recognition performance. They may need to learn to appreciate the alien social and emotional repertoire, in the same way that we need training in emotional valence detection in dogs. Hence, social learning may play a significant role in social cognition in human–dog interactions.

3.6 Overall Conclusion

Humans and other animals have different communicative systems. Our signal repertoire is quantitively and qualitatively different from other animals: humans have more words than other animals have calls or gestures, and we can communicate more than our direct needs. We learn our words differently than other animals do (using theory of mind). We have syntactic categories linked to our signals, and consider the minds of others when communicating, whereas only chimpanzees show some evidence of the latter. In principle, we can make any kind of sentence (obeying the rules of a particular language) of any length, whereas other animals seem to be constrained by a two-signal boundary for combining signals, and if larger sequences are produced, these contain redundant information. More precisely, our sentences have the properties of recursiveness, hierarchical structure and compositionality, whereas other animals seem to have weak compositionality at best: combining two signals into a sequence that is qualitatively different from the sum of the two constituting signals. The two-signal boundary hypothesis (Zuberbühler, 2019, 2020, 2021), however, needs more investigation.

Despite having different communication systems or differences in biological constitution, there is enough overlap between us and other animals to make meaningful communication possible in principle, even though this requires an effort on both sides. For example, in human–dog interaction, social learning is necessary for dogs and training is necessary for humans to understand each other's specific emotional valances.

3.7 Learning Outcomes

Source: Illustrated by Loïc Maréchal adapted from istockphotos.

Chapter Summary

- The central theme of this chapter is what communication is, how humans communicate with each other, how other animals communicate with conspecifics, and where our communicative abilities overlap.
- The human signal repertoire is quantitively and qualitatively different from other animals. For example, we have more words than other animals have signals, and we use a theory of mind when learning or using these signals, whereas – apart from the chimpanzee – other animals do not.
- Human sentences have the properties of recursiveness, hierarchical structure and compositionality, whereas other animals only have weak compositionality: combining two signals into a sequence, where the meaning of the combined signal is qualitatively different from the meaning of the two signals separately.
- There is enough overlap between us and other animals to make meaningful communication possible. As shown in relation to human–dog interaction, visual and auditory cues in both species may help or hinder communication in face perception, identity perception and emotion perception.

Check Your Understanding

Multiple Choice Questions

Question 1: Chomsky's main idea about how human and animal communication differ from each other is that:

- Answer a: humans have consciousness, and animals do not, and therefore we can consciously create sentences, whereas animals cannot.
- Answer b: humans have the unique ability to produce words, whereas animals can only produce calls or songs, which have limited meaning.
- Answer c: humans have knowledge of syntax, whereas other animals do not.

Question 2: Chimpanzees are probably the only animals apart from humans who:

- Answer a: can express themselves creatively.
- Answer b: can express intention using a theory of mind.
- Answer c: can produce both calls and gestures like we do.

Question 3: Dogs and humans read emotions in each other's faces:

- Answer a: in the same way.
- Answer b: according to the basic emotions: happy, sad and angry.
- Answer c: in different ways.

Open Questions

1. This chapter considers human–animal communication. In what way or to what extent can humans and animals understand each other?
2. Design an experiment to investigate whether dogs can understand certain prepositions (e.g. in, on, and at), or certain secondary emotions (e.g. love, guilt, shame and jealousy).

4 Bioethics and Human–Animal Interaction

Ben Mepham

4.1 Introduction

Bioethics is a branch of philosophy with one metaphorical foot in abstruse theory (e.g. where it is concerned with the concept of animal rights and the nature of non-human animal consciousness), and another in the appropriate human relationships with animals which for so long have been used as sources of food, means of traction and transportation, objects of scientific research, guards, guides, companions and, from prehistory, as beings hunted, worshipped and sacrificed. In addition, with a growing appreciation that animals often have unacknowledged cultures, skills, understanding and social interdependencies, the need for bioethical reassessment has become incontestable. Viewing issues from our almost inevitably human perspective, the chapter addresses a number of ethical dilemmas by use of the author's Ethical Matrix, a conceptual tool adapted from theories employed effectively in human biomedical ethics. In this way, the mutuality of impacts of people, animals and the biosphere is viewed through an ethical prism. In sum, the aims of this chapter are to inform readers of the nature of bioethics and to equip them with the skills for addressing the ways we should appropriately respect the interests of non-human animals.

Other chapters explore the interactions which occur between people and non-human animals in different contexts (e.g. agriculture, as pets and in the wild) and from a range of perspectives (such as scientific, cultural and legal): almost without exception, ethical concerns underlie all such interactions (Although *ethics* and *morals* are defined slightly differently, in this chapter they are used as synonyms). In the current context, many questions that are popularly classed as 'ethical' relate to animal welfare, which, as in food production systems and biomedical research, often have impacts on non-human animals' physical and/or mental well-being. However, from a philosophical perspective, the ethical remit is much wider and operates at a deeper level of enquiry. Like all branches of philosophy, ethics is replete with theories which their proponents consider the most valid in addressing the concerns that fall within its remit. So, a practical objective of bioethics is to analyse the ways competing theories seek to reach sound judgements on the value of the range of options under consideration, examples of which are given below.

What might now be apparent to the reader is that bioethical enquiry does not aim to provide *the* correct judgement on an ethical dilemma, but to enable the choice of that (or those) judgement/s that a person of integrity and with serious intent (or a group, say a committee, of such individuals) would consider intellectually sound. In addition, although it is more likely, but by no means inevitable, that people living in a similar cultural environment might hold similar opinions, across the wide range of cultural perspectives that have been adopted on a global scale, the disparities between different cultures are often quite substantial (see Chapter 6). In sum, despite the meticulous and pedantic reasoning that characterises philosophers' theorising, it is

DOI: 10.4324/9781003221753-4

rarely the case that the crucial ethical question (i.e. *What should I do in these circumstances?*) can be answered unequivocally by sincere people throughout the world.

4.2 Theories of Ethics

The two most commonly employed theories of ethics are *utilitarianism* and *deontology*, although many of who use them in justifying their beliefs or actions may not realise that they are doing so. In that both theories aim to produce good behaviour (in the ethical sense), they merit some discussion of their respective pros and cons because their approaches are quite different. Both are based on principles, but of very different kinds.

4.2.1 Deontology

This theory places a particular emphasis on the relationship between one's duty and ethical behaviour. The theory was described most definitively by Immanuel Kant (1724–1804), arguably the most important European philosopher since ancient Greece. According to an authority on his philosophy:

> His writings on ethics are marked by an unswerving commitment to human freedom, to the dignity of man, and to the view that moral obligation derives neither from God, nor from human authorities and communities, nor from the preferences or desires of human agents, but from reason.
>
> (O'Neill, 1993, p. 175)

His answer to the central question *What ought I to do?* is enshrined in the concept of *categorical imperatives*, of which typical examples are: *do not harm, do tell the truth, do not steal, do show respect to others.* Accordingly, one should act *only* on the principle: *Do as you would be done by.* From this brief summary, it doubtless appears that an uncompromising sense of moral rectitude permeates strict Kantianism – despite the fact that for Kant religious insights played no part in moral reasoning.

However, some snags with Kant's approach became apparent, as when conflicting outcomes result from attempting to obey two categorical imperatives at once. For example, when a doctor's truthful statement to a patient that her death is imminent causes additional distress – a case of truth *versus* harm. Then there is Kant's concentration on *goodwill*, which he claimed is the principal concern because any *adverse* consequences of one's good intentions are always unpredictable.

4.2.2 Utilitarianism

Deontology might come across as a rather dour recipe for an ethical life – but for most philosophers since ancient times, the principal aim of ethics is to achieve a state of *happiness*, although that word is open to numerous definitions and nuances. For example, for Aristotle in ancient Greece, the word *eudaimonia,* often loosely equated with happiness, is perhaps best interpreted in terms of 'flourishing' or 'prosperity'.

While such a state of *happiness* is commonly considered in personal terms, its generalisation to large groups of people – or humanity as a whole – was the basis of the ethical principle of *utilitarianism* advanced by the British philosophers Jeremy Bentham (1748–1832) and John Stuart Mill (1806–1873). As Bentham put it: nature has placed mankind under the governance of two sovereign masters, *pain* and *pleasure* (Goodin, 1993). He believed that these were the two motivators and goals of all human action, and morally right action aimed to maximise the 'pleasure' and simultaneously minimise the 'pain' that it is anticipated would result from alternative courses of action.

For Bentham, happiness was entirely subjective, so that enjoying a simple children's game was of equal value with appreciating the arts as music and poetry – and, since everybody could play the game, it might be said to be *more* morally worthy than elitist pleasures derived from the arts because they were appreciated by far fewer people. Mill, although supporting the principle that happiness is the main aim of moral goodness, differed from Bentham in believing that some pleasures are *better* than others (Goodin, 1993), for example, that listening to a Beethoven symphony is a higher quality experience than watching a Walt Disney cartoon.

Moreover, although as noted, deontological categorical imperatives do not take consequences into account, utilitarianism is highly dependent on the accuracy of the predictions and on which particular agents are factored into what amount to *cost/benefit* calculations. These are potentially major flaws of utilitarianism, for not only are predictions often mistaken, but knowing exactly who or what to include in the analysis and over what time period it should be conducted is highly conjectural. So, although there is a vast gulf between the underlying motives of deontological and utilitarian ethics, there is also much latitude in how the different theories are interpreted in reaching moral judgements (Smart & Williams, 1973).

Even so, it seems likely that, whether or not we have ever thought in such terms before, both these forms of ethical reasoning do play a part in our decisions on how to act in certain circumstances (i.e. in a rather homespun manner that is sometimes dubbed the *common morality*). That is to say, for many of us the context and scale of likely consequences often influence our ethical judgements.

4.2.3 Principlism

The realisation that both utilitarianism and deontology appeared to be significantly flawed approaches to ethical reasoning led Ross, an Oxford philosopher in the 1930s, to derive a structured way of addressing moral problems without dismissing principles altogether. His form of *principlism* recognised that rigid adherence to deontological or utilitarian principles is impractical, and that it is more reasonable to regard principles as of a prima facie nature (i.e. *at first appearance*), thus allowing a stronger case to take precedence over a weaker one (Ross, 1930).

Subsequently, two ethicists in the United States, Beauchamp and Childress (1979), adopted this approach in exploring the complex interaction of factors often encountered in medicine, and their seminal text *Principles of Biomedical Ethics* advocates the use of principles that lie at the core of moral reasoning in healthcare. A couple of random examples of dilemmas doctors often encounter are (i) whether to recommend invasive surgery to an elderly patient, when there are significant risks that an improved quality of life might not result and (ii) who should make decisions for treatment options for an incompetent patient who has not left an advance directive. In resolving such quandaries, they added to deontology and utilitarianism the principle of *justice* (interpreted as *fairness*, as in a fair distribution of resources and profits but also risks and responsibilities) – a line of reasoning which had been pioneered by John Rawls (1921–2002), who considered this an important criterion of ethical judgements (Rawls, 1951, 1971).

The attraction of this approach was emphasised by another medical ethicist, as follows: 'the principles provide a set of substantive moral premises on which to base reasoning in healthcare ethics and offer a trans-cultural, transnational, trans-religious, trans-philosophical framework for ethical analysis' (Gillon, 1998, p. 315). If true, such claims are highly relevant to our increasingly globalised, multicultural world.

4.3 The Ethical Matrix

About 30 years ago, I designed the Ethical Matrix as a conceptual tool, a development of the approach advanced by Beauchamp and Childress, in order to facilitate, but not to prescribe, the process of ethical reasoning applicable to human–animal interactions (for example, Mepham, 1997, 2008a, 2016). The Ethical Matrix, exemplified by Mepham (1997), was selected by Chadwick and Schroeder (2002) as one of those publications which had exerted a seminal influence in the development of the field of applied ethics during the previous 50 years. Given its acknowledged value, it is used in the rest of this chapter to examine the importance of integrating the effects of ethical impacts on different interest groups in respect of: Section 4.3.1, commercialisation of a dairy biotechnology; Section 4.3.2, keeping a companion animal in an urban environment; Section 4.3.3.1, culling badgers to limit the spread of bovine tuberculosis and Section 4.3.3.2, culling elephants in an African nature reserve.

4.3.1 Use of Bovine Somatotrophin to Increase Milk Yields of Dairy Cows

The idea of designing this conceptual tool first occurred to me when I was exploring ways in which to simultaneously take account of the interests of several distinct groups which are often involved in the resolution of ethical dilemmas – thus producing rather more complex interactions than are the main focus of the doctor/patient dyad (Mepham, 2012a, 2012b and 2013). The specific issue addressed at that time (when I was primarily involved in a research team as a biochemist) was to explore the ethical implications of injecting cattle with a hormone known to increase milk yields, which was being contemplated as an agricultural tool for increasing productivity.

The basic Ethical Matrix (EM), shown in Table 4.1, indicates the three ethical principles as interpreted by Beauchamp and Childress: viz. Utilitarian (expressed as *well-being*), Deontological (expressed as *autonomy*), and Rawlsian *fairness*. Key interest groups are *dairy farmers* (who would be free to use the technology to increase yields); *consumers* of the milk

Figure 4.1 Photo of a dairy cow.

Source: Jan-huber-sOSZl7IuX-I-unsplash.

and derived dairy products; *treated cows* (receiving regular injections of the hormone bovine somatotrophin (bST) derived from *rDNA technology*; and the *biota* (i.e. the flora and fauna of the environment) that might be affected by the use of this technology. The latter group (not all of which are probably conscious) is now generally considered to have *ethical/moral status*, and is sometimes defined as *ethically considerable* (e.g. Larrère & Larrère, 2000; Mepham, 2000).

In using the EM, the effects of fully respecting the different principles are translated into practical consequences for the interest groups concerned. For example, increased farmers' profits by using bST would respect dairy farmers' well-being (DW), whereas increased disease rates in cattle due to bST would count as infringing the principle treated dairy cows' well-being (TW). In order to perform semi-quantitative ranking of responses, failure to respect a principle might be classed as −1 (significant) or −2 (highly significant), whereas positive scores indicate respect for the principle (+1 or +2), and 0 indicates no significant impact.

Assessments of the ethically positive, negative or neutral effects of *using* bST by comparison with *not* using it were made by groups of UK individuals knowledgeable about the relevant concerns (chosen to be representative of their known *for*, *against* or *uncertain* opinions). The anonymously completed assessments were recorded at workshops, and the results collated to summarise the overall opinions given in Table 4.1.

Within the available space, it is only possible to give a very brief account of the pros and cons of analyses conducted using the EM (see Chapter 3, Mepham (2008a)) for more details, as well as free sample materials at: https://fdslive.oup.com/www.oup.com/academic/pdf/13/9780199214303.pdf). Even so, this is a good example because it illustrates how the decision on whether to license bST for commercial use, over 20 years ago, came to opposing conclusions – namely ***yes*** in the United States and ***no*** in the European Union (EU) – on grounds which, whether or not the regulatory bodies recognised it, made different assessments of the importance of respecting the specified *ethical principles*.

On this basis the grounds for approval of bST use (as in the United States) and its use being banned (as in the EU) may be summarised as follows (Box 4.1):

Table 4.1 A basic ethical matrix customised here for analysis of the use of bST in commercial dairying

Respect for	*Well-Being*	*Autonomy*	*Fairness*
Dairy Farmers	Satisfactory income and work Assessment **DW** −2 −1 0 1 2	Managerial freedom Assessment **DA** −2 −1 0 1 2	Fair trade laws Assessment **DF** −2 −1 0 1 2
Consumers	Safety and acceptability Assessment **CW** −2 −1 0 1 2	Choice Assessment **CA** −2 −1 0 1 2	Affordability Assessment **CF** −2 −1 0 1 2
Treated Dairy Cows	Welfare Assessment **TW** −2 −1 0 1 2	Behavioural freedom Assessment **TA** −2 −1 0 1 2	Intrinsic value Assessment **TF** −2 −1 0 1 2
Biota	Conservation Assessment **BW** −2 −1 0 1 2	Biodiversity Assessment **BA** −2 −1 0 1 2	Sustainability Assessment **BF** −2 −1 0 1 2

Box 4.1 Brief Ethical Analysis of bST Use in Dairying

Dairy Farmers

Well-being: some, but not all, responses, envisioned increased profits

Autonomy: US farmers have the opportunity to increase productivity, but some feel economically obliged to use bST (i.e. subject to *technological treadmill*)

Fairness: US farmers rejecting bST can only label milk '*bST-free*' at their own expense

Consumers

Well-being: an EU report suggested possible (but poorly defined) risks of consuming IGF-I (a growth factor, concentrations of which increase in milk of treated cows); but an FAO/WHO committee denied any significant health risk

Autonomy: in the United States most milk is unlabelled, denying consumers a choice on whether to purchase milk from treated cows

Fairness: there appears to be no clear evidence of any impact on milk prices

Dairy Cows

Well-being: cattle suffer increased disease rates (such as mastitis, lameness, metabolic and digestive disorders), as noted on the product label, which lists 21 possible adverse side effects; the EU banned bST largely on animal welfare grounds, but the manufacturers claim the diseases are treatable by medication

Autonomy: behaviour may be adversely affected by lameness, reduced grazing opportunities (increased concentrate feeding), and decreased fertility, for example

Fairness: arguably, excessively instrumental use of cows infringes their intrinsic value: but others claim it accords with accepted social norms

The Biota

As quantitative data are lacking, claims are largely speculative.

Claimed ***positive*** features of bST: reduced cow numbers (fewer cows needed to produce the required milk yield may (questionably) reduce environmental pollution (e.g. fertiliser use for forage growth and reduced silage run-off) and lower greenhouse gas emissions (methane is exhaled by ruminant animals).

Claimed ***negative*** features of bST: mergers in dairy industry (as non-user farmers leave the industry) will result in fewer but much larger dairy farms, thus increasing point-source pollution (e.g., excessive fertiliser use, silage run-off) and jeopardise biodiversity and sustainability by reliance on fossil fuels for fertiliser production etc. and routine veterinary medication.

- *The ethical acceptability of bST use* for those who have licensed it would probably cite the need to respect farmers' freedom to innovate; and the economic benefits to the manufacturers of bST, to the economies of countries producing it, to the farmers using it and, were prices to fall, to consumers of dairy products. Moreover, if its use led to reduced cow numbers, it might result in marginally reduced emissions of methane, thereby contributing to a reduction in global warming. This case also rests on perceptions that the welfare of treated cows is not reduced significantly (or that increased disease can be effectively treated) and that there are no risks to human safety, so that labelling is unnecessary. Job losses in the dairy industry would not be seen as an ethical issue, being merely a feature of market economies, in which competition guarantees efficient production.
- *The ethical case of those who have banned bST use* focuses on ways it infringes several ethical principles identified in the EM. Thus, authoritative reports suggest bST use substantially increases risks of painful and debilitating diseases in dairy cows, for example, foot problems, mastitis and injection site lesions, which led the *EU* Scientific Committee on Animal Health and Welfare (SCAHAW, 1999) to state bST 'should not be used in dairying'. They might also argue that bST use (i) presents risks to human safety through ingestion of increased IGF-I in milk, (ii) would reduce farmers' autonomy; (iii) would undermine consumer choice were milk and dairy products from treated cows not labelled; (iv) jeopardise public health if rejection of dairy products followed licensing of bST (as milk is a valuable source of dietary nutrients) and (v) increase environmental pollution by the intensification of dairying (Mepham, 1991, 1992).

4.3.1.1 Specific Points Needing Clarification

Data cited are those reported when the political decisions were made over 20 years ago in the United States and the EU. Since then, the availability of more data may have changed the basis of the evidence formerly advanced, either positively or negatively. However, any such amendments are irrelevant here, where the aim is not to make an actual ethical judgement, but to demonstrate the EM's use as a conceptual tool for reaching, and/or explaining, any such judgements.

4.3.1.1.1 THE ASSESSMENTS OF INDIVIDUAL CELLS

The factors in each cell of the EM which are relevant to performing an ethical analysis of the impacts of bST are of two major types. In some cases, objective *evidence* is required. For example, you would need to know what increases in milk yield were obtained when bST is injected, whether any effects on the chemical nature of the milk have implications for consumer health, and whether the welfare of the injected animals is affected. For people who see science as about 'facts', the answers to such questions might appear straightforward. However, the nature of these so-called facts, whether they were obtained reliably, and whether they are relevant to the question in hand are all matters over which there is sometimes disagreement. If the source of the data is thought to be biased (e.g. if a commercial company was relied on to produce the key data supporting their own product, or if the data were produced by a pressure group known to be ideologically opposed to the product), the evidence might be unreliable. Assessing evidence may thus entail examining different versions of the 'facts' where there is controversy.

In contrast to 'factual data', other cells of the EM require a judgement that is not dependent on the quantifiable consequences of bST use but instead concerns *values*. For example, in the pursuit of economic objectives, is it right to treat animals instrumentally by chemically altering their metabolism, or is it right to take risks with human health when appropriate scientific evidence is unavailable? People may well differ in how they assign value to matters such as welfare, risk and freedom, and while the exchange of opinions on such matters is often valuable, they are not amenable to objective measurement. It is also worth noting that the specification of the principles is, to some degree, based on subjective assessment – which is, of course, open to challenge.

4.3.1.1.2 CLARIFICATION OF CONSUMER AUTONOMY

Autonomy in this context is about liberty – being able to choose the sort of food you eat and how it was produced. In many respects, these are *citizens*' concerns and not just consumers' concerns such as accurate food labelling, because you might have legitimate views on how a food is produced irrespective of whether you consume it.

4.3.1.1.3 CLARIFICATION OF ANIMALS' INTRINSIC VALUE

The principle of fairness applied to dairy cows is specified as respect for *intrinsic value*, a term which needs further explanation. Some things (e.g. stethoscopes and taxis) are valuable because of their usefulness, and are said to have *instrumental value*. By contrast, *intrinsic value* is assigned where it is possessed irrespective of any usefulness; and most of us share the fundamental belief, stressed by Kant, that all people have intrinsic value. However, most people sometimes (and others, often) *also* have instrumental value, so that possession of the two types of value is not mutually exclusive. For example, doctors, taxi drivers and refuse collectors all perform useful tasks, making them of instrumental value. This does not raise an ethical concern if they do their jobs by choice, work in safe and congenial conditions and receive a fair income. Attributing intrinsic value to dairy cows makes the assumption that in addition to their instrumental value in providing milk and dairy products, cows are also 'subjects of a life' which we have a duty to respect. That is, they have *ethical standing*. Given all we now know about the sentience, sensibilities and even 'personalities' of cows (Young, 2017), it would be unfair to regard them *simply* as useful objects. Recent UK legislation gives official recognition to this concept (Gov.UK: www.gov.uk)

4.3.2 A Dog as a Companion Animal in an Urban Environment: Test Case for Practical Analysis

The overall procedure outlined in discussing the commercial use of bST in the United States and the ban on its use in the EU should have given sufficient insight into the ways the EM can provide explicit forms of justification that have been used to underpin political decisions. The polarisation of the judgements advanced is reflected in the different *weights* allocated to the impacts (positive and negative) on the different ethical principles. Analysis of anonymous data often reveals that it is the contrasting perceptions of the *risks involved versus* the *benefits anticipated* that distinguishes the two conclusions. However, the EM can also be useful for individuals, families or specific interest groups when faced by ethical dilemmas concerning personal HAIs. To illustrate this, readers are presented with an example for which they are invited to perform a rigorous ethical analysis and reach their own judgement in recommending an appropriate course of action.

4.3.2.1 The Scenario

Mary, a 75-year-old widow, in order to be near her forty-year-old son Jack; Jill, his wife; and their two young children, has recently moved from her home in a small provincial town to a

Figure 4.2 Photo of a Yorkshire terrier.

Source: chris-smith-vCPF8e_-JPg-unsplash.

flat in a busy major city. To 'keep her company', Jack thinks a dog, such as a Yorkshire terrier, would be ideal for Mary and also because the family could meet at weekends, allowing the children to learn about pets, which is not easy with two working parents. Mary, even when her husband was alive, had not previously kept a dog. However, in order to begin assessing Jack's idea, quite a lot more information needs to be available.

Table 4.1 shows only the basic EM for the case for bST, but for the current analysis the impact of the ***proposal*** considered requires more detailed elucidation. Table 4.2 takes this specification process much further. It should be noted that the list of issues presented in Table 4.2 is not exhaustive.

For example, Mary's recently built, but quite small, flat is on the third floor of a tower building. It is about two miles from her son's house, so that apart from weekends, more visits from the family are unlikely. A small park, where dogs can be walked on a lead, is about 200 metres from Mary's flat, but having a minor heart complaint (for which she is prescribed routine medication), she is not keen on taking too much exercise.

Mary is rather apprehensive about the proposal, but is persuaded that a dog would be a great boon in that she does not make friends easily and has felt lonely since her husband died, and also because she wants to please her son, who helps her in many ways when his busy work schedule allows.

4.3.2.2 Components of the Analysis

Because of the multiple and interacting ethical impacts of the analytical procedure, the recommendation, guided by the interpretation of the ethical principles for each interest group, is to consider each cell of the EM independently. The specification of the principles, as in all cases, has been designed to match the circumstances of each interest group, but is inevitably subjective to a degree.

Table 4.2 Ethical Matrix designed for analysis of keeping a dog in a small urban flat

Respect for	*Well-Being*	*Autonomy*	*Fairness*
Dog	1. Healthy, fulfilled life, within safe environment 2. Absence of maltreatment	1. Ability to express natural behaviour 2. Respect for dog's rights	Freedom from unwarranted restrictions (by comparison with feral relatives?)
	Assessment **DW**	Assessment **DA**	Assessment **DF**
	−2 −1 0 1 2	−2 −1 0 1 2	−2 −1 0 1 2
Keeper	Companionship and other rewards of keeping a dog	Freedom to keep a dog in suitable environment	Freedom from adverse discrimination & burdensome regulation
	Assessment **KW**	Assessment **KA**	Assessment **KF**
	−2 −1 0 1 2	−2 −1 0 1 2	−2 −1 0 1 2
Neighbours	Freedom from undesirable effects: mess, smell, noise, physical injury and threats, traffic nuisance, zoonotic disease etc.	Normal life choices unaffected by activities of dog and keeper	Freedom from adverse discrimination by comparison with non-neighbours
	Assessment **NW**	Assessment **NA**	Assessment **NF**
	−2 −1 0 1 2	−2 −1 0 1 2	−2 −1 0 1 2
BIOTA	No significant adverse effects on environmental integrity (e.g. due to carbon 'pawprint')	Freedom of wildlife and farm animals from harassment	Proportionate impact on ecological sustainability (e.g. via climate change & biodiversity)
	Assessment **BW** −2 −1 0 1 2	Assessment **BA** −2 −1 0 1 2	Assessment **BF** −2 −1 0 1 2

However, in practice even more detail is needed (e.g. pertaining to Mary's character and health, the accessibility of her flat, the age and previous life history of the dog, and Jack and Jill's ability to give adequate time to helping Mary). These considerations will doubtless arise as we consider each cell in turn. The judgement process, that involves an overall synthesis of the results recorded for each of the 12 cells, is the final step.

DW: Relevant considerations relate to:

a how much exercise the dog will get in view of Mary's heart problem and/or the possibility of hiring a dog walker.
b whether the dog receives adequate nutrition.
c appropriate living and sleeping provision.
d whether it is pampered (e.g. by overfeeding or being allowed to sleep in front of a warm fire).

Animal welfare is variously assessed according to three criteria: (i) *basic health and functioning*, which are revealed by conformity to physiological norms; (ii) *affective state* (subjective feelings), which may be appreciated by a dog previously maltreated, but which an inappropriately mollycoddled dog may relish and (iii) *natural living*, which considers welfare best assessed by *the extent to which an animal can live according to certain decision rules which were adaptive to the environment where the species evolved* (Fraser, 2008). Reliance on any one of these to the exclusion of the others is probably undesirable, and unattainable, but acknowledging them all may result in making a wise choice.

DA: This is problematical because of the difficulty of defining the natural behaviour of domesticated species, and the precise definition of animal rights in this context.

DF: This also is problematical because certain wild animals appear to have adapted to domestication to highly beneficial effect. A powerful case for this (which may well be illustrated by Jack's suggested choice of a Yorkshire terrier) has been made by Budianski (1992).

KW: Assuming that great care is taken in choosing the dog, that it is in good health and the new ownership has been officially recorded, this step should be straightforward. However, Mary's full awareness that the dog may sometimes get ill, and consequently incur the costs of veterinary fees, is an important consideration.

KA: In many countries, landlords are no longer be able to issue blanket bans on pets by default. Even so, favourable conditions (for both Mary and the dog) relating to the need to use stairs or a lift, easy accessibility to outside facilities for recreation, fresh air and toileting need to be established in advance of the dog's arrival.

KF: Notifying immediate neighbours of the impending arrival of the dog may be tactful, not least to gauge their reactions. Some may offer to help with exercising the dog.

NW: Dogs are sometimes a nuisance to neighbours and visitors and can cause alarm by barking and jumping up when confronting some people. Yorkshire terriers have a common reputation for having a loud, high-pitched bark, which is elicited when they are happy, annoyed or in need of attention – but this appears to be much influenced by training. Without knowing about Mary's neighbours, the importance of this to the analysis is uncertain. Similarly, the generally low risks of people catching diseases from dogs depend on specific circumstances, which are often listed on government websites (e.g. UK Government, n.d., a).

NA: Avoiding potential problems (e.g. of corridor congestion when taking the dog for a walk at a time when children in neighbouring flats are rushing off to school) should be considered.

NF: As for NA, attempting to fit in with neighbours' established daily routines should facilitate integration with the existing social patterns, and possibly help Mary to make friends with her new neighbours.

BW: For a small dog spending most of its time indoors, this is unlikely to present any problems.

BA: Were any visits to the country or farm sites to be made, perhaps as outings with Jack and Jill's family, care would need to be exercised to prevent sheep- or cattle-worrying.

BF: It is anticipated that there would be only insignificant impacts from such a small animal.

4.3.2.2 Conclusion

The collection of 12 scores (one per cell), each ranging from −2 to +2, is likely to give a good indication of the strength of the ethical grounds for action on Jack's proposal for Mary to have a Yorkshire terrier for company and to enrich her life. In addition, recording the opinions so that they can be revisited and reviewed, may give confidence in the decision made. However, the final judgement cannot be calculated mathematically because some, at least, of the scores are subjective. Some beliefs which depend on intuition (gut feelings) are not necessarily any less valid for that. So, despite others' good intentions, Mary may decide that, given her age and poor health, she is not confident she could provide the dog with all the necessities of a good and healthy life.

However, this does not dismiss the analysis as a waste of time, because Jack's original intention was to stimulate alternative solutions that may prove beneficial. For example, a dog of a different breed might be more suitable, or a pet animal of a different species or for Mary to become a member of an old people's group who share opportunities to make friends with a pet and have occasional outings with it. Jack may even purchase a dog when the children get older, which Mary could enjoy on her visits to their home. Exploring the ethical aspects of the original proposal may stimulate new, more satisfactory, ideas that could satisfy the underlying desire for companionship.

4.3.3 Culling Animals

Because both humans and non-human animals utilise the world's natural capital, conflicts of interest are almost inevitable. Moreover, recent sociological research, building on the study of animal behaviour, suggests the existence of *cultures* among non-human animals, thus challenging the assumption that human uniqueness provides the sole rationale for sociology as an academic discipline (Nimmo, 2012). Indeed, recently, leading philosophers whose main focus has been on ethical theory relating to humans have explored the ways in which such theories are relevant in our interactions with non-humans (e.g. Beauchamp, 2011; Frey, 2011; Hursthouse, 2011; Korsgaard, 2011).

A case is often made that, in certain circumstances, taking everything into consideration, selective killing (culling) of non-humans is an ethical requirement. To some people, this assertion may seem to challenge the very ethos of ethics: but as noted by Powys (1974, p. 13),

> if by the time we (have reached maturity) we haven't noticed what a knot of paradox and contradiction life is, and how exquisitely the good and the bad are mingled in every action we take [...] we haven't grown old to much purpose.

Is this cynicism or wisdom? The use of the EM is a way of testing the ethical validity of culling in different contexts (e.g. by comparing the practice when applied to farm and wild animals). Two examples chosen to illustrate this are (i) culling badgers, which are alleged to spread bovine tuberculosis (bTB) in herds of dairy cattle and (ii) culling elephants in an African nature reserve, the voracious appetites and other activities of which threaten ecological sustainability. Table 4.3 is a form of the EM that might be used in each case by choosing the appropriate option in certain cells. Space restrictions mean that the commentary and analysis presented are limited – but readers who have read up to this point should be able to interpret the arguments effectively. A more extensive account has been published (Mepham, 2016).

4.3.3.1 Alleged Case for Badger Culling

In England and Wales, over 40,000 cows are slaughtered per annum because they are infected with *bovine tuberculosis* (bTB), caused by *Myobacterium bovis.* Although bTB is theoretically transmissible to humans, there is virtually no risk from consuming pasteurised milk and dairy products. However, costing the UK government £500 million (2004–2014), it is claimed to be '*the most important animal policy issue in England*' (McCulloch & Reiss, 2017, p. 470).

Table 4.3 Ethical Matrix designed for analysis of culling: applicable to case of badgers/cows OR elephants in nature reserve

Respect for	*Well-Being*	*Autonomy*	*Fairness*
Animals: *badgers/cows* with, or at risk of, contracting bTB, or threatening others' interests OR *elephants*	1. Healthy life, within safe environment 2. Humane treatment in life or at slaughter Assessment **AW** −2 −1 0 1 2	1. Ability to express natural behaviour 2. Respect for animal rights Assessment **AA** −2 −1 0 1 2	Absence of unreasonable restrictions (compared with feral relatives?) Assessment **AF** −2 −1 0 1 2
Managers: e.g. government regulators, vets, farmers, OR nature reserve rangers	Human safety and confidence in efficacy of procedures employed to address the problem Assessment **MW** −2 −1 0 1 2	Freedom to act in accordance with conscience in animal management Assessment **MA** −2 −1 0 1 2	Absence of adverse discrimination & burdensome regulation Assessment **MF** −2 −1 0 1 2
Public: as citizens; food consumers in work, retirement, childhood OR at leisure	Insignificant undesirable effects (e.g. physical or zoonotic disease threats, or culturally offensive practices) Assessment **PW** −2 −1 0 1 2	Life's normal activities and choices unaffected by procedures adopted to address problem Assessment **PA** −2 −1 0 1 2	Absence of adverse discrimination by comparison with unaffected areas or countries Assessment **PF** −2 −1 0 1 2
Biota: fauna and flora in *wild environment* OR *on farms*	Insignificant adverse effects on environmental integrity and species conservation Assessment **BW** −2 −1 0 1 2	1. Freedom of farm animals or wildlife from restrictions on behaviour 2. Respect for biodiversity Assessment **BA** −2 −1 0 1 2	Insignificant impact on ecological sustainability (e.g. via climate change) Assessment **BF** −2 −1 0 1 2

Badgers may be partly responsible, but cow-to-cow transmission accounts for most of the spread, which some claim could be solved by cattle-based controls, such as improved cattle-testing regimes and more effective vaccination programmes for both cows and badgers. Even so, the UK government has persisted with the culling policy since 2012. In 2017, badger culling was practised on 9,000 km^2 of land in England, resulting in about 25,000 badgers being killed, with about 17,000 shot by marksmen and about 8,000 trapped and then shot. The latter is a much more expensive, but humane, procedure because many badgers injured by being shot endure prolonged and painful deaths.

However, critics argue that the perturbation caused by culling in one area displaces badgers to neighbouring areas, thus transferring, and possibly, exacerbating the problem. Moreover,

Figure 4.3 Photo of a European badger.

Source: vincent-van-zalinge-GvSLkDH7XdI-unsplash.

an independent expert panel (IEP) raised concerns about the humaneness of free-shooting and argued that standards needed to be improved if culling was to continue. Although the chief veterinary officer concluded in 2017 that it was unnecessary to assess future badger culls, the British Veterinary Association Council, after initially supporting the cull, raised concerns about its humaneness, and called for badgers to be killed only *after* being cage-trapped (House of Commons Library, 2018). In an open letter to the *Observer* newspaper, dated 14 October 2012, 30 eminent animal disease experts stated,

> As scientists with expertise in managing wildlife and wildlife diseases, we believe the complexities of TB transmission mean that licensed culling risks increasing cattle TB rather than reducing it. This led to a demand for the immediate abandonment of the killings.
>
> (Mepham, 2016)

In summary, referring to Table 4.3, government-supported measures, allegedly *scoring well* for **MW**, **MA** and marginally for **AW (cows)**, justified proceeding with the cull, whereas IEP experts, supported by many academic scientists and members of the public, assessed the cull as *markedly negative* for the relevant ethical principles **PW**, **PA** and **PF** as well as **AW, AA** and **AF (badgers)**.

However, in 2020, it was announced that 'as a result of a major breakthrough by government scientists, world-leading bTB vaccination trials are set to get underway in England and Wales. These trials will enable work to accelerate towards planned deployment of a cattle vaccine by 2025' (UK Government, n.d., b). It is important to note that criticism of earlier practices stimulated much-needed research and suggests that culling will soon cease to be the main method of bTB control.

4.3.3.2 Elephants in Kruger National Park

Over a century ago, elephants in Africa were virtually wiped out because they were hunted and killed for their ivory and skin. Unfortunately, the voracious appetites of elephant herds are

Figure 4.4 Photo of an African elephant.

Source: wolfgang-hasselmann-CFcGYVIKBQ0-unsplash.

highly detrimental to the stability of the ecosystem, such that the future survival of many living species, including the elephants themselves, appears to be seriously threatened. Consequently, the National Parks Board in South Africa opted to control the numbers of African elephants in the Kruger National Park (KNP) by the, highly controversial, process of culling. The aim was to balance the probable destruction of vegetation and the consequent depletion of biological diversity that high elephant densities caused. In order to keep the elephant population of the KNP to below a ceiling of 7,000, a total of over 16,000 elephants were culled between 1967 and 1994.

However, in 1994, pressure from animal rights groups led to a moratorium on future killing while the policy of culling underwent public debate and a full scientific assessment. An expert panel's conclusions suggested that the problem is extremely complex, but, crucially, that there is no compelling evidence for the need for immediate, large-scale reduction of elephant numbers in the KNP (Owen-Smith et al., 2006). The panel concluded that: (a) the previous ceiling of 7,000 should not be construed as a carrying capacity; (b) there is no benchmark against which to judge an ideal state of KNP vegetation and (c) culling alone may actually make matters worse, not least because of elephants' behavioural responses (Mepham, 2016). On the other hand, possible strategies might entail: (i) increasing the calving interval, through reduced fertility (e.g. by restricting surface water supplies); (ii) employing selective regional culling and (iii) using immuno-contraception in small, enclosed reserves (Druce, et al., 2011). Even so, other knowledgeable experts challenge these conclusions.

The complex interactions between animal and plant species, some of which are synergistic while others are predator-prey relationships, together with climatic variables and the influence of factors such as fires, floods and droughts – all produce ecological changes in states of temporal and spatial flux. These factors cause marked fluctuations in biodiversity, so that disturbances may not return to an original state for decades, if indeed they ever do so (Kruger National Park, n.d.).

From the ethical perspective argued by Regan (2006), that lays emphasis on respect for animal sentience, killing contravenes each animal's rights, but for those who prioritise ecological integrity (e.g. Callicott, 1980), individual animals' interests must be considered in a holistic context, which may, regrettably, entail *therapeutic culling*. This has been defined as that form of culling which is designed to secure the aggregate welfare of the target species across generations, the health and/or integrity of its ecosystem, or both (Varner, 2011).

With reference to Table 4.3, culling would clearly produce *strongly negative scores* for **AW**, **AA** and **AF** **(elephants)**, *but positive scores* for **BW, BA** and **BF** – and probably also, for a large proportion of the public, **PW, PA** and **PF**. However, the conclusions in terms of practical measures are by no means clear. The complexity of the interrelationships between the elephants and the various flora and fauna over time, the possibility that the use of immuno-contraception might prove a feasible strategy, and the increasingly unpredictable effects of climate change – all these complicate the decision-making process. The role of the EM in this case is perhaps limited to identifying where progress might best be made as further understanding is acquired.

4.4 Conclusion: Supplementary Perspectives for an Emerging World Order

Brevity, it is said, is the soul of wit: but it can often be the enemy of lucidity. The engaged reader might find merit in incorporating the following viewpoints into the foregoing account – thus amplifying understanding of the chapter's overall content.

1 A way in which respect for the interests of animals can best be ensured (as they would hardly ever be in the 'wild') is for us to agree to be bound by an unwritten *human–animals contract*. It would be unwritten only for the reason that the animals could not read or sign up to it. However, the same principle applies to our care for other people – such as young children and elderly people suffering from dementia. The terms of this contract (e.g. with dairy cows) would take the form: *In exchange for the benefits in regularly supplying milk for human beings, we the farmers, in a caring capacity, undertake to provide cows with a good life and a gentle death.* For some people, the flaw in the concept is that animals are almost inevitably deprived of their lives prematurely, hardly something a rational agent would agree to – although the decisive age of death might depend on the animal's health. Yet, to keep alive vast numbers of ageing, often sick, animals would present almost insurmountable economic, humanitarian and logistical challenges. In addition, since in the wild, older animals would usually die through predation, illness, serious injury or exposure, domestication under the circumstances described may still be thought, certainly in utilitarian terms, to constitute a kinder outcome (Mepham, 2008b).
2 Kaiser and Forsberg (2001) ascribe the value of the EM to several features, for example, (i) it is liberal regarding the approach to be adopted, enabling it to be read equally as a utilitarian or a deontological approach; (ii) it provides substance for ethical deliberation, guiding participants so that they do not stray into irrelevant paths; (iii) it translates abstract principles into concrete issues of direct concern to participants who may have little acquaintance with, or interest in, ethical theory per se; (iv) it facilitates extension of ethical concerns into fields benefiting from debate, such as democratic decision-making and (v) it captures the basic fact that because different stakeholders will be affected differently by a decision their ethical evaluations may well differ. The object is not to downplay these differences but to search for an optimal solution in the light of the conflicts.

3 In trying to formulate sound ethical trajectories, the sheer pace of geopolitical change to which we are now exposed was expressed graphically by the distinguished sociologist Zygmunt Bauman in the term *liquid modernity* (Bauman, 2009, pp. 110–143). Commenting on our, ultimately unattainable, pursuit of happiness, Bauman (2008, p. 20) claimed, in *The Art of Life*, that: 'Our lives, whether we know it or not, and whether we relish the news or bewail it, are works of art. To live our lives as the art of life demands, we must, just like the artist of any art, set ourselves challenges which are (at the moment of their setting, at any rate) well beyond our reach'.

Bioethics in today's world demands knowledgeable appreciation not only of its ethical, epistemological and scientific foundations but also of the social, cultural and political factors that exert such powerful influences on those who earnestly desire to live morally virtuous lives.

4.5 Learning Outcomes

Source: Illustrated by Loïc Maréchal adapted from istockphotos.

Chapter Summary

- The main ethical theories (deontology, Utilitarianism, princilism) are described, indicating their strengths and limitations.
- The manner in which these theories can be put to practical use by representing their prima facie nature in the conceptual tool called the Ethical Matrix is described – providing a means of rational decision-making in the area of human animal interaction.
- Examples are discussed regarding the application of the Ethical Matrix to facilitate the resolution of several human–non-human interactions.

Check Your Understanding

Multiple Choice Questions

Question 1: Utilitarianism is an ethical theory which advocates looking for solutions to ethical dilemmas by taking into account:

- Answer a: the greatest good for the greatest number.
- Answer b: the imperative to act as we expect to be treated by others.
- Answer c: the four ethical principles used by Beauchamp and Childress.

Question 2: The Ethical Matrix is a tool for addressing ethical issues.

- Answer a: based on universal principles.
- Answer b: in relation to humans, animals and the environment they live in.
- Answer c: by using objective measures so that it is possible to make a calculated decision.

Question 3: An important prerequisite for using the Ethical Matrix is:

- Answer a: to determine how all the relevant parties are affected by an ethical decision.
- Answer b: to know all the ethical theories discussed in this chapter.
- Answer c: to have an idea of the possible outcomes of the ethical decision-making process.

Open Questions

1. Some pet shops sell electric dog collars, which can be used to administer a shock to a dog for the purpose of training or to modify the dog's behaviour in different ways. Using the Ethical Matrix, is it possible to provide strong grounds in deciding whether this is a sound ethical practice?
2. What are the ethical considerations surrounding human–animal welfare, and how can these considerations be balanced with competing interests and priorities, such as economic development, cultural traditions, and conservation goals?

5 Human–Animal Welfare

The Interconnectedness of Human Well-Being and Animal Welfare

Laëtitia Maréchal, Ana Maria Barcelos, Justine Cole and Meagan King

5.1 Introduction

What is it about human–animal interactions (HAIs) that affects both humans and animals? Can we really influence each other's welfare? The 'One Welfare' concept promotes the idea that animal welfare, human well-being and environmental sustainability are interconnected, and thus, influence one another (Pinillos et al., 2016, 2018; Figure 5.1). One Welfare is an extension of the 'One Health' concept as it considers mental health as well as physical health (see Chapter 1, this volume). Indeed, the term 'welfare' can be used for all animals, including humans. In this chapter, welfare is defined as the physical and psychological health of an individual. In humans, as welfare is often associated with economic and social factors, it is worth noting that the term 'well-being' is commonly used when the focus is on physical and mental health only. Therefore, in this chapter the terms welfare and well-being will be used interchangeably for humans.

This chapter uses the One Welfare framework to highlight the human dimension of the intricate relationships between humans and other animals. While environmental aspects under the One Welfare concept remain important components, this chapter will focus on the human and animal elements. More specifically, this chapter will first introduce the impacts of interacting with animals on human well-being, followed by the second section on the human factors that impact animal welfare, keeping in mind that these relationships are often not unidirectional. Each section will present the factors described in the following framework (Figure 5.2), when possible, including examples from different types of HAIs from domesticated species (including farm and companion animals) and wildlife.

Figure 5.1 (a) The QR code of the One Welfare video (www.youtube.com/watch?v=lZHlknHGGL0).

Source: See One Welfare website for more information: www.onewelfareworld.org/.

DOI: 10.4324/9781003221753-5

Figure 5.1 (b) One Welfare umbrella outcomes.

Source: See One Welfare website for more information: www.onewelfareworld.org/.

5.2 Animal-Related Factors Influencing Human Welfare/Well-Being

Human welfare or well-being can be influenced by interactions with other animals. However, the association between HAIs and human well-being is complex because many factors influence this relationship. In this section, we explore the main animal-related factors that have been shown to affect human welfare such as attitudes towards animals, the types of interactions, animal-related activities and animal end of life. We provide examples from companion, farm and wild animals, although not all factors have been explored in each HAI, often highlighting the need for further research.

5.2.1 Attitudes towards Animals

Attitudes towards animals (i.e. the way in which a person thinks and feels about animals) have important implications for the outcomes of HAI, including their impacts on human well-being. For instance, interacting with a dog might be perceived as relaxing for people who have positive associations with dogs, but stressful for others who are fearful of dogs. In addition, attitudes towards animals can influence how people behave towards animals, consequently affecting the HAI and the outcome for human and animal welfare. For example, people fearful of dogs

Figure 5.2 The human–animal welfare framework developed in this chapter.

might behave unpredictably (i.e. sudden movements), and such behaviours have been shown to increase the likelihood of being bitten by dogs (Wright, 1996). In addition, research suggests that strong attachments to pets are associated with prosocial behaviour and a positive attitude towards animals, which can result in positive physical and emotional well-being in children (Hawkins et al., 2021). Moreover, farmers' greater empathy towards animals was associated with better quality of human–livestock interactions, resulting in greater occupational well-being (Leon et al., 2020). Furthermore, attitudes towards animals can mediate the impact of wildlife on humans' welfare. For example, positive attitudes towards wild animals can increase conservation efforts and animal tolerance, and so result in human–wildlife coexistence (Kansky et al., 2014; Mogomotsi et al., 2020). However, negative attitudes towards animals might exacerbate conflicting situations with wildlife, and lead to decreased human and animal welfare.

Attitudes towards animals can be influenced by various interconnected factors including: (1) individual characteristics such as personality, mood, beliefs and ethics; (2) social aspects such as culture, income, education, values, age and gender; (3) information available associated with exposure to and consumption of different forms of media; (4) previous experience with animals and (5) the characteristics of the animal involved (e.g. species preference/likeness, types of use or the degree of biological or behavioural similarity to humans) (Batt, 2009; Herzog et al., 2015). Overall, these factors are inherently driven by the socio-economic and cultural contexts in which each person lives, which in turn influence the information about animals to which a person is exposed. To date, most research has been conducted on specific participant groups with little socio-economic and cultural diversity. Therefore, more research is needed to better understand people's attitudes towards animals (Su & Martens, 2017), and how such attitudes affect human welfare.

5.2.2 Types of Interactions with Other Animals

5.2.2.1 Animal Ownership

Research analysing people's narratives on how the animal they own impacts on their lives (usually referred to as qualitative research) consistently highlight the predominant benefits of animal ownership (Barcelos et al., 2020; Love, 2021; Maharaj & Haney, 2015). In contrast, studies focused on numbers and statistical significances, mainly cross-sectional (i.e. data collected at one specific point in time such as surveys) in this field, show mixed results about the effect of animals on owners' mental health (Brooks et al., 2018; Islam & Towell, 2013; Rodriguez et al., 2021). It is not yet clear whether simply having an animal, as compared to not having one, has a significant beneficial effect on human well-being (e.g. happiness, depression, anxiety, stress: Rodriguez et al., 2021). Studies with more robust methodology such as randomised controlled trials assessing the effects of animal acquisition could help disentangle this. However, such studies are hard to execute and scarce in the literature. In randomised controlled trials, for instance, similar participants are randomly allocated to groups – the intervention group, where in this case participants would acquire an animal, and the control group(s), where they would not acquire an animal – to assess the efficacy of the intervention (National Institute for Health and Care Excellence – NICE). Adding an animal to the household is no simple task and requires thoughtful consideration from their future owners. Thus, a common challenge in randomised controlled studies on animal ownership is to find similar individuals willing to be randomised to one of these groups and act accordingly (intervention or control) for a specific duration predetermined by the researcher.

A few longitudinal studies showed that acquiring an animal reduces self-reported symptoms of depression (Pereira & Fonte, 2018; Potter et al., 2019), which strengthens the evidence in favour of the psychotherapeutic effect of animals. However, these studies have one or more of the following limitations: small sample size, gender bias (usually more female participants), lack of a control, volunteering bias (i.e. people who are willing to acquire an animal might be more open to change) and lack of participant randomisation. These limitations hinder the robustness and generalisability of the findings. Furthermore, solid evidence is even more scarce in relation to the impact of captive or farm animals on human mental health, despite suggestions that interacting with these animals may also be beneficial for human well-being (e.g. lowering blood pressure and stress levels, decreasing depression and facilitating relaxation; Cox & Gaston, 2016; Pedersen et al., 2012; Sakagami & Ohta, 2010). For example, interviews with 480 farmers in France, the Netherlands and Sweden revealed that farmers can develop a strong connection with their animals (Bock et al., 2007). However, while some farmers see their production animals as their own pets, referring to them as friends and family members, others perceive them simply as tools for production and money. This variation is partly triggered by the type of animal kept, as owners of cows tended to report a stronger emotional bond with their animals than those dealing with pigs and poultry (Bock et al., 2007). Different levels of closeness not only affect the human–animal relationship but also the outcomes (i.e. the psychological benefits of the interaction). This is the underlying principle of the 'One Welfare' approach that there are 'interrelationships between animal welfare, human well-being, and the physical and social environment' (Pinillos, 2018, p. 28).

In addition to the impact of animal ownership on psychological health, there are many physical health benefits and risks associated with owning animals. For example, dog ownership has been shown to increase exercise for a person, which is linked to improved physical health (e.g. reduction in cardiovascular disease) (Mubanga et al., 2017; see Section 5.2.3). Owning a pet can also contribute to a reduced risk of asthma and allergic rhinitis in children exposed to pet allergens during the first year of life (Ownby et al., 2002). However, there are also some risks associated with owning animals. For instance, dog bites have been recognised as a serious public health risk (WHO, 2018), which are mainly directed towards owners and children (Overall & Love, 2001). It is important to note that it is very challenging to get an accurate estimate of dog bite incidents, as most bites are not reported (Overall & Love, 2001). Zoonoses (pathogen transmission from animal to human) are also an important risk. For example, cats can transmit a parasite called *Toxoplasma gondii* to humans though their faeces, which can have adverse effects on human foetuses and immunocompromised individuals (Dabritz & Conrad, 2010). The popularity of exotic pets (i.e. non-native to a region or non-domesticated animals) has also shown an increase in envenomation (i.e. exposure to a poison or toxin resulting from a bite or sting from an animal such as a snake, scorpion, spider, insect or fish) (Ng et al., 2018).

5.2.2.2 Interacting with Wildlife

While wildlife encounters through recreational activities are thought to elicit positive emotional responses in humans, offering physical and psychological benefits such as a reduction in physiological stress levels (Sumner & Goodenough, 2020), sharing space with wildlife can present both positive and negative impacts on human mental health. Wildlife can present an exploitable resource (e.g. through hunting) and bring economic and psychological well-being benefits (i.e. biophilia hypothesis: Wilson, 1984; Chapter 2, this volume). However, sharing space with wildlife can also create some conflicting situations (e.g. economic loss or danger), which can negatively affect people's mental health (Bond & Mkutu, 2018; Chapter 11, this volume). In addition, interacting with wildlife presents some risks for humans' physical health, ranging from injury to death. For example, a recent study has shown how human casualties outweighed property losses in conflicts linked to wildlife in India (Gulati et al., 2021), highlighting the importance of taking this factor into consideration when managing human–wildlife interactions. Recently, debates around human–wildlife interactions (e.g. illegal wildlife trade or deforestation) being linked to higher zoonosis risk, have raised serious concerns for public health (Fagre et al., 2022). Therefore, human–wildlife interactions can directly affect physical health (i.e. injury, death, zoonoses) and also indirectly influence psychological health as seen by the negative outcomes of the lockdowns on mental health during the Covid-19 pandemic (Fiorillo & Gorwood, 2020).

5.2.2.3 Working with Animals

Animal care professionals (e.g. veterinarians, technicians, animal control officers) often show high levels of compassion satisfaction, described as the positive feeling that comes from helping others, but also compassion fatigue, defined as a combination of burnout (i.e. exhaustion, depression and frustration towards one's working environment), and secondary traumatic stress associated with animal illness or death (Polachek & Wallace, 2018, see animal death section below). This is often called the paradox of compassionate work (Polachek & Wallace, 2018). For example, compassion satisfaction for animal rescue staff decreased when burnout and secondary traumatic stress increased (Murphy & Daly, 2020). Animal care professionals also self-report high levels of compassion fatigue, greater than those reported for human service professionals like healthcare workers and police officers (Hill et al., 2020). Veterinarians scored significantly higher in terms of burnout, compassion fatigue, anxiety and depression, while scoring lower in terms of their resilience (Perret et al., 2020). Suicide ideation and planning are more prevalent in veterinarians, and these individuals are 2.7 times more likely to commit suicide than the general population (Perret et al., 2020; Volk et al., 2022). There may be even more mental health and burnout concerns for veterinary support staff compared to veterinarians (Volk et al., 2022). In addition, animal care professionals are at risk from many occupational hazards including trauma (e.g. scratches, bites or kicks from animals), zoonoses, drugs, vaccines, anaesthetic agents, pesticides, insecticides and allergens from animals (Jeyaretnam et al., 2000).

Farming is another occupation which is highly stressful, both physically and mentally. Stress, anxiety and depression among farmers (as well as their spouses and partners; Hounsome et al., 2012) are generally higher than for non-farmers and the general population (Jones-Bitton et al., 2020). There are also differences within farming populations, with farmers who raise livestock showing greater rates of stress-related symptoms and suicide than farmers who only grow crops (Kanamori & Kondo, 2020; Chapter 10, this volume). There is currently very little data comparing farmers across commodity groups, but there are additional stressors such as public perception, animal activists and disease outbreaks (Donham et al., 2016). Like veterinarians, farmers are also at risk from many occupational hazards including trauma, zoonoses (e.g. swine or avian flu), pesticides, insecticides and allergens from animals (Drudi, 2000; Żukiewicz-Sobczak et al., 2013).

5.2.3 Animal-Related Activities: Costs and Benefits

The biopsychosocial model could help explain how animals influence human well-being (Gee et al., 2021). According to this model, the 'biological', 'psychological' and 'social' aspects of an individual's life interact with each other. Thus, improvements in mental health from interactions with animals are not only restricted to the psychological domain, but they can also originate from changes in the social element (e.g. social support from an animal, more social interactions with other people due to the animal's social lubricant effect) and the biological element (e.g. better physical health due to exercise with the animal; changes in hormonal levels from interactions with the animal).

A novel approach to understanding how animals affect human well-being (and influence the three elements of the biopsychosocial model) is to focus on the activities between humans and animals and the specific impact they have on mental health (Barcelos et al., 2020). An animal-related activity is defined as any situation or event that happens due to the existence of the animal in the person's life, such as petting the animal, going to the vet and buying the animal toys (Barcelos et al., 2020). At least 58 different animal-related activities have been reported by dog owners living in the United Kingdom (Barcelos et al., 2020) and Brazil (Corrêa et al., 2021), and 67 activities have been reported by cat owners in the United Kingdom (Ravenscroft et al., 2021), each having a specific impact on human well-being. The exact animal-related activities with specific human well-being outcomes in interactions with other animal species (e.g. farm and wild animals) and in different contexts (e.g. slaughterhouse, animal shelters or wildlife tourism), remain mostly unexplored, but are also expected to have specific impacts on human well-being.

A qualitative study about the role of animals on suicidality illustrates the advantages of focusing on specific animal-related activities within the human–animal relationship (Love, 2021). Love (2021) found that animals' behavioural issues and health issues were risk factors for owner suicide, while the companionship of the animal and the obligation to care for it were some of the protective factors against suicide. Therefore, relationships with animals are not universally positive or negative for an individual's mental health. Human–animal relationships are complex, with pros and cons that must be carefully assessed before recommending an animal acquisition as part of a psychotherapeutic intervention.

Positive activities for human well-being include having tactile interactions with the animal, experiencing the animal's presence, and walking, training, feeding, being greeted by, and playing with the animal. Higher frequency of pet presence has been associated with better self-reported well-being in owners, such as greater happiness and relaxation, and lower feelings of depression and anxiety (Bennett et al., 2015). Walking animals, particularly dogs, has been linked not only with physical benefits but also with psychological ones, such as an increase in social interactions with other humans (McNicholas & Collis, 2000) and protection against loneliness (Carr et al., 2021). Caring for animals often provides a routine, purpose and structure in people's lives, and can also represent a way of connecting with nature and relaxing, as reported by people who feed birds regularly (Cox & Gaston, 2016). Finally, in zoos, caring for animals allows professionals to develop bonds with the animals, which is described as emotionally rewarding, joyful and beneficial for increasing sense of responsibility, resulting in higher job satisfaction (Hosey & Melfi, 2012).

Although the literature tends to highlight how beneficial animals are, some animal-related activities can be detrimental for human well-being. This includes dealing with animal sickness, injury or death (see section on animal end of life); experiencing animal behavioural problems (e.g. aggression, house soiling); failing to fulfil the animal's needs and paying animal-related bills. Animal behavioural problems can cause a lot of stress for humans, particularly for owners and animal carers. In a study with Australian veterinarians, for example, nearly half of the participants reported being bitten or scratched by a cat (67%) or a dog (48%) in the previous 12 months (Fritschi et al., 2006). Finally, caring for a sick animal is commonly a burden and source of stress to those involved in the animal's care (Spitznagel et al., 2017).

5.2.4 Death/End of Animal Life

5.2.4.1 The Loss of an Animal: Grief Manifestation

The death of a beloved animal is a difficult experience in human–animal relationships (Lavorgna & Hutton, 2019). For many owners, an animal's death can also represent the end of constant companionship, social support and unconditional love, and may cause an enormous disruption in the routine of the person left behind (e.g. the structure of the day shaped around animal feeding breaks is lost). Grief is defined as 'emotional, cognitive, functional and behavioural responses to [a] death' (Zisook & Shear, 2009). After losing an animal, people often feel depressed, lonely, anxious, upset, empty and angry (Archer & Winchester, 1994). There are five stages of grief recognised in both the human and human–animal literature (Rujoiu & Rujoiu, 2013): denial (disbelief of loss), anger (at themselves, at the veterinarian or at the deceased animal), bargaining (wish that situations from the past could be altered), depression and acceptance. These stages do not necessarily occur consecutively, and some people may even skip stages or repeat some of them over time (Ventura et al., 2021). Although most pet owners experience at least one symptom of grief after the death of a pet, their loss is not always understood or recognised by others, which is referred to as 'disenfranchised grief' (Corr, 1999). The grief for non-pet animals may be even less understood, as most scientific studies and media attention are directed at pet owners. It is important to seek grief support for pet bereavement. However, one should refrain from suggesting the acquisition of another animal, as this could be seen as diminishing the value of the bereaved animal and such an acquisition could interfere with the healing process of grieving (Lagoni, 2011).

5.2.4.2 The Caring–Killing Paradox

Having to perform euthanasia, or even being exposed to euthanasia, can put animal shelter and laboratory staff at a greater risk of burnout and compassion fatigue, termed 'The Caring-Killing Paradox' (Andrukonis & Protopopova, 2020; LaFollette et al., 2020). Burnout and compassion fatigue may therefore negatively impact the ability for laboratory and shelter staff to effectively manage the animals in their care. Having to make the choice of which animal to euthanise in shelters (Andrukonis & Protopopova, 2020) as well as feeling a lack of control or choice over euthanasia and using physical methods such as cervical dislocation, penetrating captive bolt and blunt force trauma (LaFollette et al., 2020) are additional stressors. Therefore, although shelter staff have higher job satisfaction in shelters with more positive outcomes (i.e. higher percentage of adoption), they may also be subject to greater traumatic stress when animals are euthanised (Andrukonis & Protopopova, 2020).

5.2.4.3 Slaughtering, Culling and Hunting

Slaughtering animals for meat or culling purposes can also have negative impacts on the mental health of the individuals performing those tasks. Government officials and veterinarians involved in the culling of farm animals to control disease outbreaks reported high depression rates, greater post-traumatic stress disorder scores and insufficient mental health care (Park et al., 2020). Further, the physical and psychological well-being of slaughterhouse workers is worse than for many other occupations, even when controlling for the 'low prestige' and 'high dirtiness' rating of the job, showing that intentionally killing animals has a unique consequence (Baran et al., 2016). This may cause perpetration-induced traumatic stress as well as nightmares, alcohol abuse and inner conflict about the person's work. For example, the effects on human health were particularly devastating during the Covid-19 pandemic in crowded meat processing plants in the United States, where workers are often members of ethnic minority groups and/or migrant communities without sick leave or health insurance (Marchant-Forde & Boyle, 2020). When plants were forced to shut down, this had systemic effects on human, animal and environmental health, especially for pig and poultry farms that had to cull and dispose of millions of animals.

Baran et al. (2016) highlight three key differences between slaughterhouse work and hunting wild animals: (1) slaughterhouse work involves the routinised and systematic killing of countless animals, (2) it is 'up close and personal' and (3) the animals being slaughtered have been reared by, and may even trust, humans. Given the many reasons why people may have to slaughter or cull animals, there are unique and complex repercussions for those individuals for which additional mental health supports are likely needed.

Finally, research has mostly reported the positive impacts of hunting wild animals on human welfare. For example, wildlife hunting can provide a needed food resource, or a means of protecting material possessions (e.g. crops, house), and thus, this activity presents some positive impacts on human welfare (Strong & Silva, 2020). Furthermore, recreational hunting of wild animals has been associated with nature-based benefits such as stress relief, exercise and camaraderie (Campbell & McKay, 2009). However, some research also suggests that hunting, and especially poaching, can have dramatic impacts on biodiversity, resulting in decreased economic potential and, thus, potentially negatively affecting human well-being (Fisher et al., 2011).

5.3 Human Factors Influencing Animal Welfare

The concept of 'One Welfare' has also encouraged researchers to acknowledge that human factors can be major determinants of animal welfare. In this section, we examine the six components of Cole and Fraser's (2018) framework which outlines the human dimension of animal welfare. Although these factors were first described for human–livestock interactions (Hemsworth et al., 1993), many are applicable to interactions with domesticated animals, including companion and research animals, as well as zoo animals and wildlife. However, it is worth remembering that close interactions with wildlife raise additional ethical and conservation concerns, which must be taken into consideration alongside animal welfare issues.

5.3.1 Human–Animal Interactions

HAIs can either impair or enhance the welfare of animals, and the quantity and quality of these interactions will determine the development of a positive, a negative or a neutral human–animal relationship (HAR). For example, negative interactions such as shouting, smacking or using an electric prod have been shown to consistently induce fear in animals, impairing growth and productivity in many farm animal species, including pigs, sheep and cattle (Chapter 10, this volume). Additionally, aversive training techniques, which aim to reduce undesirable behaviours by vocal or physical punishments can jeopardise both the physical and the mental health of dogs (Deldalle & Gaunet, 2014; Guilherme Fernandes et al., 2017; Ziv, 2017).

In wild and captive animals kept in zoos, there is evidence to suggest that negative interactions may also augment animals' fear of humans and negatively impact their welfare. For example, aversive noises made by zookeepers, including the rattling of padlocks or the jangling of keys, have been shown to increase avoidance and fearfulness in species such as maned wolves and cheetahs (Carlstead, 2009). Moreover, tourist presence and interactions can also alter wild animals' activities, primarily by increasing avoidance behaviour, vigilance and by reducing resting, social and feeding behaviours (Constantine et al., 2004). There is evidence that human presence also increases physiological indicators of stress across taxa, including birds: yellow-eyed penguins (Ellenberg et al., 2007), and mammals: Barbary macaques (Maréchal et al., 2011).

In contrast, positive interactions with domesticated animals, such as gentle stroking and soft vocalisations, can often reduce behavioural and physiological stress responses and augment positive HARs. For example, dairy cows which experienced positive interactions such as gentle stroking before an aversive veterinary procedure had lower heart rates, kicked less and displayed less restless behaviour during the examination, indicating reduced fear and behavioural stress responses (Schmied et al., 2010). In dogs, reward-based training methods that use pleasant stimuli such as vocal praise, social contact or play have also been associated with fewer stress-related behaviours (Deldalle & Gaunet, 2014), and fewer long-term behavioural problems (Guilherme Fernandes et al., 2017). Positive HAIs can also be a form of enrichment for zoo animals as they have been shown to increase species-specific behaviours and reduce abnormal behaviours (Claxton, 2011). For example, in chimpanzees and common marmosets, just 10–20 minutes of talking, grooming, feeding or playing with their keepers increased play and grooming between conspecifics and reduced abnormal behaviours such as regurgitation (Baker, 2004; Manciocco et al., 2009). Finally, in laboratory rats, positive human contact which mimics conspecific social interactions, such as tickling, can ameliorate some of the stress associated with handling and experimental procedures (Cloutier et al., 2012).

Overall, the evidence suggests that positive HAIs have the potential to improve HARs, reduce fear of humans, help mitigate stress and greatly improve animal welfare in a wide range of species. Moreover, while negative experiences may not be completely avoidable (e.g. veterinary procedures or transport), increasing the number and frequency of positive HAIs requires few additional resources and provides a major opportunity for improving HARs and animal welfare. While the studies examined here suggest that gentle handling is most effective for domesticated animals, the effects of positive human contact can also be species-specific. For instance, handling interactions are not recommended with wild animals, and passive human presence or visual contact may be sufficient to reduce fear in many species. Further research is therefore needed to determine species-specific recommendations, including the type, frequency and duration of positive interactions required to enhance animal welfare.

5.3.2 Consistency and Familiarity

The predictability of HAIs strongly affects an animal's anticipation, perception and response to human interactions (Rault et al., 2020). For example, unpredictable human interactions, such as random electric shocks when interacting with a human, have been shown to elicit a strong stress response in both pigs and dogs (Hemsworth et al., 1987; Schalke et al., 2007). Consistent and repeated exposure to similar HAIs, however, often leads to habituation by the animal, in which they learn to predict and respond in the same manner to similar human interactions (Williamson & Feistner, 2011). Caretaker familiarity can also affect an animal's sense of predictability, impacting their welfare. For instance, shelter cats that were positively and consistently handled by a few researchers had lower stress levels and higher adoption rates than cats housed in typical shelter conditions, which involved inconsistent handling by various staff (Gourkow & Fraser, 2006). Thus, with repeated exposure to interactions with the same human, animals may also become accustomed to that specific human's presence, leading to familiarisation (Williamson & Feistner, 2011). Some animals, such as laboratory rats, even display a preference for familiar humans, and will consistently choose a familiar human over an unfamiliar human, even after just five brief bonding sessions (Burgdorf & Panksepp, 2001; Davis et al., 1997). Thus, the consistency of HAIs, as well as the familiarity of caretakers, are key factors influencing an animal's sense of predictability and welfare.

Consistent positive interactions with familiar keepers have also been demonstrated to reduce the reaction of zoo animals to visitors, suggesting that some species may generalise their experience with familiar humans to unfamiliar people. While it is species-dependent, most research suggests that captive wild animals generally find the presence of the public to be disruptive to normal behaviours (i.e. the 'visitor effect') (Hosey, 2000; Hosey, 2008). For instance, while Western lowland gorillas and orangutans commonly seek proximity to and engage in affiliative behaviours with familiar keepers, they are more likely to hide or act aggressively towards unfamiliar visitors (Smith, 2014). However, additional positive reinforcement training with a familiar keeper, for example, significantly reduced animal-initiated interactions with zoo visitors in Abyssinian colobus monkeys (Melfi & Thomas, 2005). Nevertheless, while habituation and familiarisation can reduce the stress responses of wild animals, it is also important to consider the ethical concerns and consequences of reducing wild animals' fear of humans, such as increasing the risk of poaching, disease susceptibility, intra-species behavioural modifications, or human–animal conflict with nearby human communities (Green & Gabriel, 2020).

5.3.3 Recognition of Animal Individuality

Individual animal characteristics, such as age, intelligence, genetics or animal personality may also modulate an individual animal's perception of, and reaction to, HAIs. For example, animal personality, defined as the distinct pattern of behaviour that is characteristic of an individual and consistent over time, has been recognised in a wide range of companion, domesticated and wild species (Finkemeier et al., 2018; Gosling & John, 1999). Of the five factors used to describe human personality, three of them (extraversion, neuroticism and agreeableness) have also been described in numerous animal studies (Gosling & John, 1999). Thus, differences in animal personality may influence individuals' responses to HAIs. For example, Afromontane chacma baboons' individual differences in flight distance indicated that habituation to humans was modulated by their personality (Allan et al., 2020). Moreover, cheetahs whose personalities were scored as significantly more 'tense-fearful' by their keepers were also found to be significantly less likely to breed, suggesting that these individuals might have more difficulty acclimating to captivity (Wielebnowski, 1999). While the role of individual factors such as personality requires significantly more research, studies like these provide preliminary evidence that individual differences do have implications for the animal's health and welfare.

Individual recognition of animals is therefore necessary for caretakers to notice subtle changes that may indicate compromises to an animal's health and welfare as well as identify animals in need of additional management or care. There is currently very limited research on the effects of individual recognition of animals, but one survey of UK dairy farm managers found that it had a significant effect on the welfare and productivity of their dairy herds (Bertenshaw & Rowlinson, 2009). On dairy farms where individual cows were called by name, cows were easier to approach and milk yield was significantly greater than on farms where animals were not individually recognised (Bertenshaw & Rowlinson, 2009). Individual recognition might therefore contribute to a more positive human–animal relationship and reduce stress to the animal by encouraging caretakers to tailor their interactions to each animal's individual personality (Cole & Fraser, 2018).

5.3.4 Attitudes and Personality

A person's behaviour towards animals is also influenced by the attitudes and beliefs they hold about the animals (Herzog et al., 2015; Chapter 10, this volume). Serpell (2004) describes human attitudes towards other animals as a spectrum (i.e. from negative to positive) which are based on two main motivational considerations: affect (i.e. individual affective or emotional responses to animals) and utility (i.e. people's perception of animals' instrumental value) (also see Chapter 6, this volume). Based on this model, attitudes towards animals scoring low in affect and utility are likely to be associated with negative behaviours towards these animals, and vice versa. For example, animals considered and treated as pests such as rats and cockroaches would score low in affect and utility. However, many human and animal characteristics influence people's attitudes towards animals, as seen in Section 5.2.1. Attitudes are therefore an important determinant of human behaviour, and a key factor in building positive HARs and ensuring animal welfare.

A caretaker's personality also influences their behaviour towards animals and can impact their HARs and animal welfare. Dairy cows, for example, were found to be less reluctant to enter the milking parlour and were less restless during milking in the presence of stockpeople described as 'confident introverts' (Seabrook, 1984). More recent research has also shown several links between the 'Big 5' human personality traits (extraversion, neuroticism, agreeableness, openness, conscientiousness) and the behaviour, health and welfare of companion animals. For example, dogs with owners who scored higher for 'neuroticism' displayed higher rates of aggression towards strangers, their owners and other dogs (Gobbo & Zupan, 2020). Similarly, cat owners who scored higher for 'neuroticism' were more likely to report their cats being overweight, more aggressive and/or possessing stress-related sickness behaviours such as urinary infections, vomiting or diarrhoea (Finka et al., 2019). In contrast, owners who scored higher for 'conscientiousness' reported their cats displaying less anxious, avoidant or aggressive behaviours, and those who scored higher for 'agreeableness' were more likely to indicate their cats were within a normal, healthy weight (Finka et al., 2019). Although more research is necessary to determine how human personality traits impact animal welfare, it has been suggested that caretakers who score higher for 'neuroticism' may increase animals' stress levels and exacerbate behavioural problems by possessing more unpredictable styles of caring (Gobbo & Zupan, 2020). Together, evidence overall suggests a relationship between owner personality traits, HAI and animal welfare outcomes.

5.3.5 Knowledge and Experience

Experience and knowledge of the behaviour, health and welfare requirements of animals is essential for all animal caretakers, as it is fundamental for ensuring animal welfare (Hemsworth & Coleman, 2011). It is also important that animal caretakers can recognise deviations from normal health or behaviour, and seek and implement appropriate measures to address welfare concerns (Hemsworth & Coleman, 2011). For example, animal transporters reported patience, knowledge and experience with sheep as key factors in reducing fear in sheep and easing handling (Burnard et al., 2015). Among zookeepers, an animal's response to their keeper (taken to be an indicator of good husbandry skills and a positive HAR) also tended to be quicker with keepers who had extensive experience and knowledge of the species in their care (Ward & Melfi, 2015). It is worth noting, however, that long-term keeper experience (up to 35 years) induced greater fear of humans in cheetahs, maned wolves, black rhinoceroses and great hornbills. It is thought that this might be due to poor habits or outdated information, which may cause those with long-term experience to perform more aversive behaviours (Carlstead, 2009).

A lack of knowledge is thought to be one of the primary issues for animal welfare in pets. For example, owners' lack of education and experience were suggested to be linked to higher behavioural problems, obesity and ill-health due to poor breeding practices (e.g. flat-face breeds) in dogs (Philpotts et al., 2019). A lack of knowledge and experience are also the main issues associated with improper care in exotic pets from owners and animal care professionals (Warwick et al., 2014, 2018).

Training programmes designed to improve animal caretakers' welfare-related knowledge have also shown promising results. For example, laboratory staff training in animal welfare awareness and the 3Rs (replacement, refinement, reduction; Russell & Bruch, 1959; for information see: https://nc3rs.org.uk/the-3rs) resulted in enhanced animal welfare knowledge and a greater understanding of the ethical implications and implementation of the 3Rs (Franco & Olsson, 2014). Similarly, among abattoir employees, those who participated in a training course on animal welfare expressed more confidence in their ability to improve the welfare of animals in their care (Descovich et al., 2019). This improvement, however, was only present in groups where both the instructor and the participants were trained in person in a classroom setting (as opposed to remote learning). These results suggest that programmes to improve caretakers' welfare-related knowledge can be effective, but that consideration must be given to the target demographic and mode of information delivery (Descovich et al., 2019). Although limited, research on the knowledge and experience of animal caretakers does underscore the importance of providing proper training on animal behaviour, handling and welfare, and that the education of animal caretakers may be a low-cost means of improving animal welfare. However, such training or education programmes have shown mixed results for pet owners (Philpotts et al., 2019, Warwick et al., 2018), suggesting that alternative solutions are needed to improve pet welfare, including changes in legislation to better control practices (e.g. proposition to ban flat-face dog breeds in the United Kingdom, or the licencing of the ownership of certain exotic pets).

5.3.6 Caretaker Well-Being

As discussed above, there is compelling evidence for a reciprocal relationship between animals and their caretaker's well-being, with notable consequences for animal welfare. This reciprocal relationship is further supported by studies of dairy farmers, workers and their cattle, which found that poor occupational health, as well as self-reported levels of anxiety and work-related stress, were positively correlated with higher herd incidences of clinical illness, such as lameness and mastitis (King et al., 2021; Kolstrup & Hultgren, 2011). Moreover, many reported cases of serious neglect in farm animal care have been attributed to personal stress and the resulting deterioration of physical and mental health among farmers (Andrade & Anneberg, 2014; Devitt et al., 2015). This reiterates the need for adequate support for human caretakers, in order to improve their animals' welfare.

Instances of animal hoarding also highlight the link between caretaker well-being and animal welfare. Animal hoarding – in which owners accumulate large numbers of animals but fail to provide minimal standards of nutrition, hygiene and veterinary care (Reinisch, 2008) – has only recently been recognised as an indication of a mental disorder by the American Psychiatric Association (2013). While hoarding is considered a severe form of animal abuse and neglect under many criminal codes, the prosecution of such cases often does not adequately address the underlying mental health problems of the individual (Lockwood, 2018). For instance, some animal hoarders have reported suffering traumatic events, leaving them with the inability to foster healthy relationships with other humans (Reinisch, 2008). Thus, to identify, treat and prevent severe companion animal neglect, it is important for hoarders to receive the necessary mental health support and treatment.

In addition to psychological health, caretakers' physical health can affect animal welfare. For example, a low amount of daily exercise was associated with higher noise sensitivity and separation anxiety in dogs (Tiira & Lohi, 2015). It was also found that owners of obese dogs tend to be obese themselves (Sandøe et al., 2014), showing the interconnected relationships between human and animal welfare.

5.4 Conclusion

Although our understanding of HAIs in laboratories, zoos and with wildlife is growing, most literature is focused on domesticated animals – primarily farm and companion animals. Overall, understanding the impacts of animals on human well-being is complex, and this chapter highlights the importance of applying more rigorous research design to examine the effects of each animal-related activity individually. The research presented in this chapter also provides considerable support that the human factors exert a considerable influence on animal welfare. In conclusion, there is compelling evidence that HAIs exert reciprocal effects on both human well-being and animal welfare, making the One Welfare framework a key concept for further research.

5.5 Learning Outcomes

Source: Illustrated by Loïc Maréchal adapted from istockphotos.

Chapter Summary

- Human–animal interactions and relationships exert reciprocal effects on both human well-being and animal welfare, both positive and/or negative (One Welfare framework).
- Understanding how interacting with animals affects human well-being is complex because it is often challenging to disentangle the different factors involved in such interactions, i.e. attitude towards animals, types of interactions and animal-related activities.
- More robust evidence is needed for both the positive and the negative impact of interacting with animals on human well-being, especially for professional caretakers.
- The main human factors that influence animal welfare include: (1) the nature of the human–animal interactions and relationships (i.e. positive, negative or neutral), (2) the consistency and familiarity of animal caretakers, (3) the treatment of animals as individuals, (4) the attitudes and personalities of caretakers, (5) caretaker knowledge and experience and (6) caretaker well-being.

Check Your Understanding

Multiple Choice Questions

Question 1: What are the five main factors influencing people's attitudes towards animals?

- Answer a: Individual factors, social factors, information, experience, animal characteristics.
- Answer b: Individual factors, personality, culture, information, experience.
- Answer c: Individual factors, culture, information, economic factors, experience.

Question 2: What should be recommended for an animal owner who is grieving the loss of their animal?

- Answer a: To avoid memorialisation of the pet (e.g. funeral ceremony).
- Answer b: To acquire a new animal to help in the healing process.
- Answer c: To seek assistance from a mental health professional well equipped to deal with animal bereavement.

Question 3: How can an animal become more familiar with a human caretaker?

- Answer a: Through positive interactions with as many caretakers as possible.
- Answer b: Through repeated positive interactions with a few specific carers for long periods of time.
- Answer c: Through short positive interactions with a few specific carers.

Open Questions

1. Describe the overall effects (positive and/or negative) of animal interactions and activities on human well-being and draw a diagram illustrating these effects. On another sheet of paper, describe and draw a diagram depicting how human factors impact animal welfare. For both diagrams, consider a variety of interactions, among both domesticated and wild species.
2. How can promoting human–animal welfare benefit both humans and animals, and what are some practical strategies for improving the welfare of both humans and animals in various contexts, such as in agriculture, wildlife conservation and pet ownership?

6 Cross-Cultural Variation in Human–Animal Interaction

Shelly Volsche, Jennifer Wathan, Naeem Abbas, Laura Kavata Kimwele, Giovanna Capponi, Ricardo R. Ontillera-Sánchez, Eva Zoubek and Melanie Ramasawmy

6.1 Introduction

The interactions between humans and other species are intrinsically tied to different sociocultural understandings. The same animal can acquire different significance depending on the context, and sometimes multiple ways of defining and interacting with animals can coexist within the same geographical and cultural sphere. For example, chickens can be simultaneously kept as a pet, used for sport or religious practices or slaughtered for their meat.

In this chapter, we explore cultural diversity in human–animal interaction (HAI) as demonstrated by three different animals, dogs, equids and chickens. These animal species were chosen because each represents an animal which millions of people and industries rely on, and provides substantial, varied and meaningful contributions to society. These representations are linked to cosmology, industry, food systems, and for many, daily life. The goal of this chapter is to provide a sampling of these relationships and inform the perception of HAIs as multifaceted and involving more than pet keeping or companionship.

6.2 Human–Dog Interaction

6.2.1 Introduction

For many, the term 'dog' conjures an image of a pampered, often purebred dog, living in the home, likely co-sleeping with its owners and going for regular walks, to the park or to dog sporting events. Relatedly, the practice of pet parenting (Volsche, 2018, 2021) is on the rise, as a growing number of people are choosing not to have children and to invest in deeply bonded relationships with their dogs, and cats. Yet even in countries with a large population of pet dogs (e.g. United States, United Kingdom), relationships between humans and dogs vary significantly. Therapy and service dogs are trained to perform atypical tasks and control their own impulses so as not to be a health hazard in 'no dog' spaces. Other dogs work with the military and police, or as livestock guardians, spending their most crucial socialisation periods in training (Coppinger & Coppinger, 2001; Graham & Gosling, 2009). Livestock guarding dogs also contribute to the large number of free-ranging dogs, and an estimated 3.3 million dogs enter US shelters each year (ASPCA, 2021). These data stand in stark contrast to the pet-friendly image conveyed in mass media and advertising, and the varied roles of dogs in cross-cultural contexts. And this is all within one country.

DOI: 10.4324/9781003221753-6

6.2.2 Understanding Cross-Cultural Variables

Serpell (2011) suggests a biaxial model that intersects affection and utility to consider both the archetype 'dog' and individual dogs within a culture. On the Y-axis, affective/emotional value, people may perceive dogs with fear and loathing at the bottom, and with love and sympathy at the top. As dogs move up the Y-axis, they may become pampered, included as family or valued as crucial to the family's mode of production (e.g. herding dogs in pastoral communities are often beloved). As dogs move down the Y-axis, they may become ostracised as dangerous, loathed as pests or viewed as filthy. The X-axis, utility/instrumental value, ranges from beneficial to detrimental to human interests. As dogs move towards beneficial, they may again be valued for their help in subsistence strategies (e.g. hunting), as useful in therapeutic settings or simply for companionship. As dogs move towards detrimental, they may attack herd animals, destroy furniture or carry disease or parasites (Figure 6.1). While variations within communities also influence exactly where dogs may fall on the spectrum, it is important to keep in mind that most dogs, even in places where pet keeping is common, are more likely to fall into the low-affect and high-utility quadrant as beasts of burden, free-ranging populations or otherwise non-NATIVE (Neutered, Alimented, Trained, Isolated, Vaccinated and Engineered: Koster, 2021).

A true exploration of dogs around the world would require an entire book unto itself. However, this cursory survey of the varied perceptions and purposes of dogs serves as an opportunity to begin seeing dogs in a new light. These high-level perspectives are tied together by a common

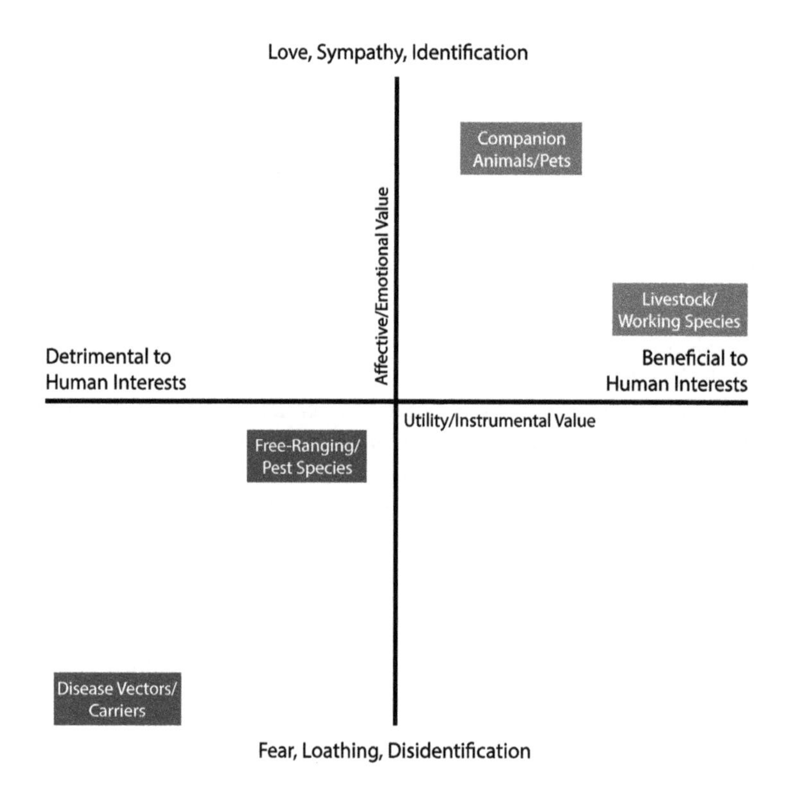

Figure 6.1 The biaxial model described by Serpell (2011) with examples of where various dogs/species may lie.

thread – they all demonstrate the flexibility, process and constantly changing roles dogs play in many parts of the world.

6.2.3 Indigenous Perspectives

In *How Forests Think*, Eduardo Kohn (2013) details the relationship between dogs and the Runa people of the Upper Amazon. Valued as hunting partners and village protectors, dogs are observed closely when dreaming. When a dog barks in its sleep, this is perceived to indicate how a dog will bark during the next day's hunt. This suggests dogs have a knowable inner life, like men. In this way, dogs and men are deeply connected, such that a dangerous or distracted dog is akin to a dangerous or distracted man. When this occurs, the people may administer the liquid from an agouti's (*Dasyprocta punctata*, a small, South American rodent) bile duct by holding the dog and tying its mouth closed, then dripping the liquid into the snout. It is believed that the dogs are fighting evil spirits, and those that survive the procedure are victorious against the spirits, while those who die are not.

Though the story of Runa dogs seems cruel, this treatment of dogs in smaller-scale societies and foraging communities is not uncommon. In cruel, challenging environments, with constrained resources, life often negates the ability or cultural norms that allow for pampered pets. Rather, dogs work hard for resources, assist foragers in hunting and protection, and scavenge for their own calories. They may be free-ranging or community 'owned', but are usually malnourished, abused and infested with parasites. This can be surprising for people from complex, post-industrial societies. Suzman (2014) documents his experience being 'adopted by Dog' among the Ju/'hoansi Bushman. Suzman (2014) tells of burying Dog, sharing this anecdote of the tribe's response: 'They had found my habits of feeding and petting Dog odd enough, but the idea of giving it a human burial just seemed absurd'.

This is not to say Indigenous or foraging people view all dogs with such dismissal. On occasion, a particularly good hunter or companion may gain favour. When queried about an egalitarian hunter-gatherer society in 2014, Crittenden acknowledged that, like most small-scale societies, dogs were often treated as 'just animals'. However, a particular member of the camp appeared to have an affinity for his dogs, regarding them with affection and speaking of love, even feeding them on occasion. She mentioned being aware of how the dogs were noticeably calmer around this one individual. This relationship contrasts with the more common behaviour of shooing and throwing things at the non-human animals in the area and speaks to the individual variability of human–dog relationships.

6.2.4 India in Transition

India poses an interesting space in which relationships between people and dogs are rapidly changing. In much of rural India, dogs still roam freely in villages (Arluke & Atema, 2017), and may be feared due to disease (Menezes, 2008) or for the threat they pose to livestock or wildlife (Home et al., 2018). Yet as India urbanises, indie (or local free-ranging) dogs fill spaces in alleys and around trash heaps. These dogs are neither owned nor removed, and instead are viewed by many urban Indians as members of the community who share this space. Interestingly, a growing number of urban Indians now report strong attachments to pets in the home (Volsche et al., 2021).

Particularly in urban centres, a distinction remains between free-ranging dogs as undesirable pets and purebred dogs, often British breeds, as desirable. Though this dichotomy exists in many parts of the world, it creates a particularly stark contrast given the history of British colonialism in India, with a bias towards the social capital received by owning one of these British breeds. However, this is changing. Volsche et al. (2019) surveyed urban college students in Bangalore, India, with interesting results. Nearly half of the students surveyed acknowledged they would like to have a dog in the future, and almost 40% stated they would prefer an indie (local stray, 19.1%) or did not have a preference between indie dogs and purebreds (18.8%). This is a marked change, as many of the students surveyed acknowledged that their perceptions of indie dogs were indeed different from those of their parents and grandparents.

Regardless, India's economic growth and cultural changes provide keen insight into an array of human–dog relationships. From unowned, free-ranging village dogs, to communally cared for indie dogs, to the rising class of pampered pets, India is clearly transitioning. As a result, owned dogs experience a suite of treatments that is becoming more like their Western counterparts. Dogs are now reported to sleep indoors, even co-sleeping with humans; eat high-quality commercial or home-made food; and receive quality veterinary care. This speaks to the ways global information exchange introduces new ideas that can be incorporated into existing cultural norms.

6.2.5 Asian Meat Markets

China, Thailand and, to a lesser extent, Vietnam, are known for their dog meat markets. These markets range in quantity of dogs slaughtered and sold, the state in which dogs are kept before consumption, and the means of acquiring dogs. Cultural relativism and a One Health model can be combined to bridge social practice and health outcomes for humans, dogs and other species sold as meat throughout the world – particularly since meat consumption in general has been connected to infectious disease and biodiversity loss (Morand, 2020).

Though pet dog keeping has emerged in China in recent decades, the Yulin Dog Meat Festival is held each year in Yulin, Guangxi, China, as part of summer solstice celebrations. The Chinese Animal Protection Network (capn-online.info) was formed in 2004 in part to work against this festival, along with the World Animal Foundation and other international organisations. It was cancelled in 2021, though it is uncertain if this was due to protests or to the Covid-19 pandemic.

South Korea's history is a bit more complex. Podberscek (2009) surveyed South Koreans about the practice of eating dog and cat meat in their culture. While many individuals no longer practise this tradition, they supported others' right to do so. This suggests a cultural tension as pet keeping began to grow. Koreans felt that external pressures to ban dog meat were an attack on their culture, while at the same time, a growing number of them were choosing to discontinue the practice. In 2018, South Korea officially ruled that it was no longer legal to raise, kill or sell dogs strictly for the consumption for meat (Brady, 2018) – though this ruling did not fully ban consumption of dog meat. This change is likely due to the emergence of a new, cosmopolitan generation.

Regardless, as evidenced by the Covid-19 pandemic, the ways in which humans relate to animals has immediate, sometimes tragic implications for everyone. This has led to a growing community of scholars, medical professionals, conservationists and others calling for a One Health/One Welfare approach. Reducing the number of open-air markets, including those that sell dog meat, is a step in that direction.

6.2.6 Conclusion

Reductions in total fertility (average number of births per woman), urbanisation, increased education and a search for personal fulfilment over strict adherence to norms are all markers of the second demographic transition (SDT; Lesthaeghe, 2014). It appears that as cultures transition through the SDT, an emergence in pet keeping occurs, and with it, a concern for animal welfare. This phenomenon has been noted in India, with more work in Asian cultures taking place (Volsche, 2018, 2021; Volsche et al., 2021). Given time, it is likely that an increased number of dogs will become pampered pets. However, as diverse as human cultures are, there is also a great chance there will always be dogs which are not NATIVE.

6.3 Human–Equid Interaction

Horses, donkeys and their hybrids – mules or hinnys, herein referred to as mules – are domestic species of *Equus*. There are approximately 116 million equids globally (Table 6.1): 57 million horses, 50.5 million donkeys and 7.9 million mules (FAO, n.d.). Africa has the largest equid population, followed by Asia, South America, Central America and the Caribbean, North America, Europe and Oceania (Table 6.1). An estimated 100 million of these are working animals (Allan, 2021). Recognised by the UN Committee on Food Security in 2016 as 'working livestock', horses, donkeys and mules are critical to the livelihoods and resilience of millions of people throughout the world (Brooke, 2016).

Working equids directly and indirectly support income generation across sectors including agriculture, forestry, construction, brick kilns, tourism, mining and public transport. They are predominantly draft animals, and although there can be an assumption that 'working horses can be deemed obsolete, relics of a time before the mid-20th century's mechanisation of farming' (Clark, 2021), many cultures continue to utilise the traction force of working equids. Even where mechanisation happens, it does not always replace working livestock. For example, Brazil has become more industrialised, but food security is still heavily reliant on the small-scale farming sector which use equids for traction (Graeub et al., 2016).

Through their ability to transport goods and people, working equids support people from the individual household to the infrastructure of major cities (Figure 6.2). In many countries, equids are critical to household tasks, like accessing community water and transporting children to school. More broadly, working equids support the waste management systems of cities and towns across the world. In Karachi, Pakistan's largest city with over 20 million people and more than 14,000 tonnes of solid waste produced daily, residents reported that there would be a huge garbage build-up if donkey carts were not available (Shah et al., 2019).

Working equids play a central role in agriculture and food production, including in soil tillage, production of manure fertiliser, transport of goods to market and transport of livestock provisions. In some areas they may directly produce agricultural products such as meat and milk. In parts of Senegal, families produce 78% more groundnuts, 46% more maize and 45%

Table 6.1 Equid populations by global region in 2019

Region	*Equid Population*	*Donkeys*	*Horses*	*Mules*
Africa	38,884,998	30,640,119	7,397,922	846,957
Asia	28,414,784	13,234,286	13,925,579	1,254,919
Central America & Caribbean	16,615,869	3,746,598	9,118,134	3,751,137
Europe	4,818,518	121,247	4,696,333	938
North America	11,158,518	51,977	11,102,540	4,001
Oceania	354,353	8,924	345,429	-
South America	17,315,235	2,780,421	12,455,788	2,079,026

Source: Allan, 2021.

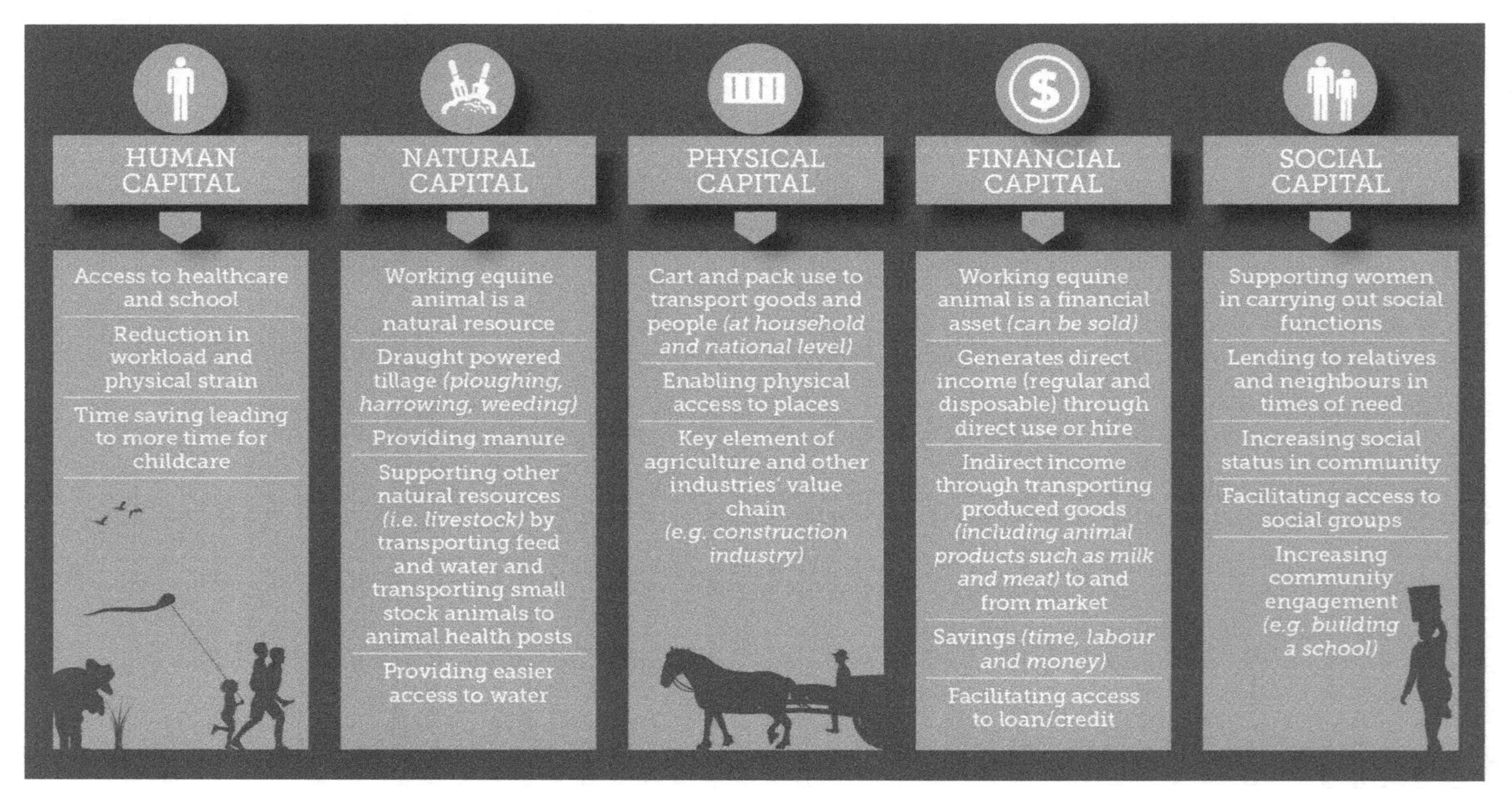

Figure 6.2 How working equine animals contribute to people's livelihoods.

Source: Image © Brooke 2014.

more millet if they have a donkey (Diop & Fadiga, 2018), and in Burkina Faso farmers would anticipate over a 50% loss in most cultivated products if they were to lose their working equids (Brooke West Africa, 2021).

In many societies, women are responsible for a large variety of labour-intensive activities, which when combined with household and community responsibilities makes a substantial workload. Working equids support women to fulfil these responsibilities, freeing up time for women to pursue education and self-development they might otherwise miss. Working equids can also raise women's status in the community and provide them with opportunities to make their voices heard and access loan and business opportunities. When asked to rank their livestock, 17 out of 22 groups of women interviewed across Ethiopia, Kenya, India and Pakistan put their horses, mules and donkeys in the first position: all groups in India and Kenya ranked working equids first (Valette, 2014).

In addition to the huge variety of ways that working equids contribute to the daily lives of people, they are also relied upon in emergency events (such as natural disasters), where working animals are often the only form of transport that can access affected areas. During the Covid-19 pandemic, the Dutch Committee for Afghanistan (DCA), a member of the Zoonotic Diseases Committee, accessed remote villages to deliver guidance on social distancing and hygiene and PPE kits using working equids (Brooke, 2020). Working animals are used post-disaster to cope with a lack of infrastructure and can support community resilience through adaptation to new environmental conditions, the supporting of livelihoods, and diversification of income streams (Weldegebriel & Amphune, 2017). In refugee camps, it has been documented that in the absence of access to banks, animals are used as a form of storing financial capital, their presence providing security and peace of mind (Alshawawreh, 2018). Finally, they can provide a more environmentally sustainable way of working, reducing vulnerability to future disaster events (Rodrigues et al., 2017).

In the following sections, we investigate the HAIs of working equids and those who depend upon them in two case studies: an arid area of Kenya and the brick kiln industry of Southeast Asia.

6.3.1 Case Study: Turkana County – Kenya

Turkana County is an arid area of North-West Kenya bordering Uganda, South Sudan and Ethiopia. The donkey population was estimated at 558,187 in 2009 but has declined substantially to 116,806 in 2019 (Kenyan National Bureau of Statistics, 2019). In this pastoralist area, donkeys are used when searching for food and water and transporting personal belongings, young children, the elderly and young livestock (goats and sheep). Some of these areas are prone to insecurity (e.g. livestock raids), and donkeys are used to carry the injured and sick to health centres; women particularly rely on their donkeys to transport household goods as they move to safer areas. Kakuma Refugee Camp, which at the end of July 2020 hosted 196,666 registered refugees and asylum seekers (unhcr.org/ke/kakuma-refugee-camp), is based in Turkana County and donkeys are used by the government and NGOs for relief food distribution activities.

Due to the different ways donkeys are used, various animal welfare issues are commonly observed. These include:

- Wounds and injuries: nasal septum damage due to handling and restraint practices; wither, spine and belly wounds caused by use of unsuitable harness and padding materials; cuts and mutilations, especially during inter-communal livestock raids.
- Heat stress: due to the need to travel long distances during migration in very high temperatures with no shade or water.
- Disease: animal health services are inaccessible in most areas in Turkana County, and even where they are available, owners often do not seek treatment (Brooke East Africa, 2016).

The Turkana community traditionally consumes donkey meat and milk as food and medicine to treat diseases including dystocia, skin diseases, tuberculosis, ulcers and pneumonia. Donkeys are also used in payment of dowry for a woman when she is getting married, and donkey hides are used to make clothing and bedding, both for use and for sale. Donkeys are slaughtered by trauma to the head with a wooden mallet or by piercing the heart using a spear, making slaughter methods another welfare issue.

The dramatic decline in the donkey population in Turkana County is attributed to the international donkey skin trade. Growing demand for *ejiao* – a gelatin produced from donkey skin and used in traditional Chinese medicine and beauty products – has led to demand for donkey skin that China cannot supply internally, and China has turned to global markets (The Donkey Sanctuary, 2019). Donkey slaughter and export in Kenya has led to a large number of donkeys entering the skin trade, many as victims of theft – in 2017, around 60 donkeys a week were stolen from owners in Kenya and across the borders. In other cases, donkeys are sold by their owners for a short-term cash injection (Carder et al., 2019). The growth in demand has increased the price of donkeys in Kenya by 50%–100% since 2016, and one study found that half of donkey users could not raise the capital to purchase replacement stock after losing their donkey (Maichomo et al., 2019). The increased, often unregulated, trade has already contributed to transboundary disease outbreaks, and poor disposal of donkey carcasses – often left to rot in the open – threatens public health (Skippen et al., 2021), making this a crisis for people, animals and the environment.

6.3.2 Case Study: The Brick Kilns of Southeast Asia

Since 6000 BCE kilns have been used to bake clay bricks, clay pots, ceramics and tiles, and bricks and tiles continue to be used globally in construction industries. South Asia's brick kiln industry still uses traditional technology to produce bricks – a process that involves thousands of people and animals: over 150,000 brick kilns employ approximately 500,000 working horses, donkeys and mules (Brooke, 2019b).

This industry hosts many people from marginalised sections of society working in conditions that are risky, hazardous and exploitative for both people and animals. Most of the people are working under bonded labour. Encompassed by the International Labour Organisation's definition of forced labour, debt bondage occurs when a person is forced to work to pay off a debt; they receive little or no pay – most or all the money they earn goes to pay off their loan, and they have no control over their debt (ILO, n.d.). This leaves people unable to negotiate government-prescribed minimum wages and other social benefits. In addition, brick kilns have severe environmental effects, including harvesting of topsoil and harmful emissions (see Figure 6.3; example Box 6.1).

Box 6.1 Example of the Working Conditions of Anwar, a Brick Kiln Worker

Anwar lives in Ajmal Brick kilns Hyderabad, Pakistan. Anwar, his family and their donkey work approximately 10–12 hours a day. They are very dependent on the income generated by their working donkey. Anwar reports that it is only because of his animal that he earns a livelihood for his family of six people. He earns about $6.4 USD per day through transportation of bricks.

THE BRICK KILN INDUSTRY IN SOUTH ASIA

ONE HEALTH ONE WELFARE CYCLE IN BRICK KILNS

HUMAN LABOUR

(SDGs 8 & 10)

The brick making industry in South Asia relies on the manual labour of around 5 million workers; 68% of them are estimated to be in forced labour and approximately 19% of the workers are under 18 years of age. A health study on child labour estimated that brick kilns engage about 1.7 million children in India, at least 500,000 in Pakistan, 110,000 in Bangladesh, and 30,000 in Nepal.

60% of kiln workers in Pakistan live below the poverty line; 80% of workers have no running water at home, 60% have no latrine facilities and 82% do not have proper drainage.

Poor working conditions, with a lack of health and safety measures creates a vulnerable environment for workers. Brick kiln pollution is responsible for 750 premature deaths annually in Bangladesh and people on the kilns breathe in fumes from toxic emissions.

The bulk of the annual income is derived from goods transport activities (>50%). 15% of annual household energy needs. Mostly received as payment for the transportation of harvest from fields to sorting area/ storage. It is very common to sell surplus in kind payments.

ENVIRONMENT

(SDG 13)

100,000 brick kilns may use up to 400 million tons of good quality top soil each year causing soil erosion and environmental degradation. Largest stationary sources of BC (black carbon) emissions internationally include brick kilns, coke ovens, and industrial boiler, accounting for 20% of total worldwide BC emissions worldwide. The industry also burns around 35–40million tonnes of coal per year, emitting carbon dioxide and sulphur, contributing to air pollution.

The main raw materials used are soil and coal. A study from Bihar (Development for Alternatives) showed that as much as 92% of the soil used for production in the state was procured from agricultural land, and if kept in agricultural production, this land could have produced 7000 tonnes of rice, enough to keep 110,000 people with food and grain.

(SDGs 1,2,4 & 8)

Donkeys, mules and horses are commonly used in the construction industry and can work up to 10 hours a day providing a means of livelihoods to their owners, plus additional work at home.

In Indian brick kilns, working animals can generate 80% of equine owning family's income. Owner livelihoods are therefore dependent on healthy animals.

The working and living conditions for these animals are very harsh at brick kilns with less access to health and other services. Overloading, overwork, heat stress, harness lesions, poor body condition and inappropriate handling are all examples of welfare compromise.

Limited national policies in place to protect working animals and their owners, with even weaker levels of enforcement.

Figure 6.3 The relationship between people, animals and the environment in the brick kiln industry in South Asia (Brooke, 2019).

Source: Image © Brooke 2019.

The living conditions of the workers are similar across the region. The workers live in small compounds on the brick kiln site, in buildings that are often poorly constructed, with crowded conditions – usually a family in one room. Equid owners mostly keep their animals at home, and all family members interact with them in different ways. Men, women and children provide drinking water, feed and clean stables. A wide range of approaches towards the animals is seen – some people have a functional view towards the animals, but in other cases the working animals are often considered like family members for their owners and if someone's animal has died, their relatives send them condolences and visit their home.

However, there is limited information on the donkeys, mules and horses in the kilns as they are virtually invisible within the literature and published evidence. The data that do exist are largely from animal welfare organisations working in the brick kilns. This shows the animals are typically in very poor welfare, leading to poor health and reduced life spans. This in turn leads to a loss of the ability of these animals to work and provide these vital functions for their owners, in addition to costly veterinary/healthcare bills (ILO, Brooke, & The Donkey Sanctuary, 2017).

6.3.3 Conclusion

Working equids represent an estimated 86% of the total global equid population (see Table 6.1); however, research output and attention are disproportionately focused on equids, specifically horses, used for leisure and sport. International development investment and poverty reduction strategies give inadequate attention to the enormous contributions that livestock provide, and equids specifically are often excluded from national livestock ministry agendas (Brooke, 2019a). Even where it is acknowledged how vital animals are to people's lives, the focus is on production animals rather than working animals (Stringer et al., 2015).

Furthermore, research about the relationship between people and their working equids often focuses on socio-economic contributions of animals to people's lives and the tangible roles that animals play. The psychological aspects of the human–animal relationship are often overlooked by research, despite one recent study finding that socio-economic functions were often spoken about concurrently with affective statements (Geiger et al., 2020). Communication between humans and animals in this context, for example, the handling of animals, is also given inadequate attention. Handling animals is essential for husbandry and healthcare; however, handling becomes a much bigger part of the daily experience with working animals. Working with equids can be very dangerous (Parkin et al., 2018), and working equid owners report that improved handling substantially reduces the stress they experience in their lives (Brooke/ESAP evaluation data). Utilising the theoretical concepts and frameworks that underpin HAI research could offer a fuller understanding of the relationships between working equids and their owners.

Keeping animals in positive health and welfare is crucial to ensuring they can lead long, healthy lives. Working equids often suffer very poor health and welfare, which has significant impacts on the people who rely upon them. Yet as we have seen in the case studies, the owners of working equids typically live in challenging environments and are providing their animals with the best standards of care that they can. Recognition of the connections between people, animals and the environment through the One Health (see above) (Mackenzie & Jeggo, 2019) and One Welfare – the concept that animal welfare depends on and influences human welfare and environmental sustainability (Pinillos, 2018) – is essential if we want to understand the full picture of HAIs and create positive change. The case studies presented here demonstrate the complex environments in which people and animals live and how important it is to understand these on local, regional and international scales.

6.4 Human–Chicken Interaction

6.4.1 Introduction

Chicken meat is the most consumed meat in the world (OECD, 2022). Poultry is also a cheap, low-input source of protein in rural areas (Alders & Pym, 2009). However, chickens have a richer history of interaction with humans than simply as sources of food. Archaeological evidence suggests that chickens were domesticated from the red junglefowl (*Gallus gallus*) at more than one site in Asia, and although the date has been disputed, this may have occurred as early as 8000 BCE. The domestic chicken then spread via Oceania to South America, and across Europe, possibly via commercial roads through the Near and Middle East (Pitt et al., 2017). Although now the focus on poultry is as a source of meat, it is thought that early adoption in many cases around the world was for other uses, including for cockfighting, for feathers, for noise and for ritual, or cosmological uses (Sykes, 2012). In this section, we will look at the different relations between chickens and humans in the contemporary world. We will explain how chickens can be ambivalent pets, gender empowerment tools, performing fighters and powerful religious symbols.

6.4.2 Keeping Chickens in the United Kingdom

In Great Britain, there has been a significant increase in the number of small-scale chicken keepers who keep chickens for non-commercial reasons in the gardens of their homes in the last 10 years (Zoubek, 2017). More than being a niche subculture, British hobby farmers constitute a profitable target for all sorts of dedicated guides, published print media, YouTube channels and social media groups aimed at sharing tips and suggestions for non-commercial chicken keeping and breeding. The attitude towards and relations with these birds create ambivalence regarding the predominant role of chickens as food.

In fact, small-scale chicken keepers have both utilitarian and emotional relationships with their chickens. Often, they refer to their chickens as pets, yet eat their eggs and at times consume their flesh. The relationship keepers develop with these birds challenges the common distinction between pet and livestock. To properly learn how to take care of their chickens, small-scale keepers go through a process of learning and sensuous enskillment (Zoubek, 2017). This involves how to properly feed the chickens, keep them clean from parasites, spot signs of illness, listen to and interpret the diversity of chicken sounds, and observe their movements, interactions and general behaviour for hours. These sensuous experiences turn people into skilled chicken keepers. Keeping chickens on a non-commercial, small-scale basis also means preparing a comfortable environment in which they can live. Keepers need to adapt their garden and backyard to the presence of chickens, choosing chicken coops or houses, erecting enclosures and creating new spaces and boundaries within outdoor and indoor property. The many different types of chicken architectural structures on the market respond not only to general notions of animal welfare, but also to the keepers' aesthetic tastes. It can be argued that the attitudes and motivations of chicken keepers seem to relate to the back-to-the-land movement, nostalgia for the countryside, and the romanticised imagery of the rural ideal. Keeping chickens becomes a way to reproduce the ideal without giving up the conveniences of modern British life.

6.4.3 Keeping Chickens in Ethiopia

In the Amhara region, in the northern highlands of Ethiopia, the association between poultry and women is reflected in production (taking place within and around other home responsibilities), as well as in language and ritual practice around childbirth. Beyond their economic use, the slaughter of chickens plays an important role in mediating relationships with the spirits that populate the Amhara landscape, including combining indigenous systems of belief with Christian and Muslim celebrations. The consumption of chickens reinforces relationships within a household, social networks, and ultimately as a form of building nationality.

In small-scale production, there has been resistance to 'improved' chickens, bred for productivity and introduced in Ethiopia since the 1950s. Practical considerations such as requirements for greater investment in specialised shelter, feeds, vaccinations and treatment are recognised to play a part in this. However, women who care for chickens also convey strong preferences for traits expressed by local breeds, including behaviour and physical characteristics considered necessary for sacrifice and consumption (Figure 6.4). As mentioned by Ramasawmy (2017, p. 251), conversations about chickens often started conversations about the changing social and economic contexts in which these relationships occur. For example, in peri-urban and urban areas around Addis Ababa (Ramasawmy et al., 2018), men show an increased interest in intensification and commercialisation of production. While women were positive about keeping chickens, and wanted to increase production, they prioritised traits that were suitable for low-input care around the home and did not require input and labour they otherwise could not afford. These findings emphasise the need for development initiatives focused on empowering women to consider the local context and who their approach will benefit.

6.4.4 Gamecocks

Gamecocks come into the picture when we talk about cockfighting. Cockfighting is suggested as one of the oldest and most widespread sports, excluding hunting (see Dundes, 1994; Smith & Daniel, 2000). It is still present in many parts of the world and enjoys considerable popularity

Figure 6.4 Illustrations of preferred physical characteristics of chickens taken to market in the Amhara region of Ethiopia.

Source: From Ramasawmy 2017.

among men, regardless of its legal status (Dundes, 1994). As Marvin (1984, p. 65) states, 'the cockfight is based on an observed fact of nature that two cocks in proximity will fight and out of this "natural fact" is created a cultural event'.

A recent ethnographic study in the Canary Islands (Spain), where the activity is legal, has shown the importance of understanding the rearing of gamecocks to fully grasp the meaning given to the fights by Canarian breeders and aficionados (enthusiastic fans of cockfighting; Ontillera-Sánchez, 2020). Aficionados watch and admire the drama of the contest, and the fighting birds' performance. In the Canary Islands, cockfighting is an indoor event where roosters, two at a time, are pitted against each other in a circular, fenced structure elevated above the floor. This pit (*la valla*) is surrounded by seating for the audience, usually composed of a few dozen people from various backgrounds, including a small proportion of women. In the Spanish archipelago, breeders can compete both in single-date championships (*campeonatos*) and weekly leagues (*contratas*). These two methods of organisation represent one of the singularities of cockfights in the Canary Islands where the relatively low importance of betting, in comparison with other regions, represents another peculiarity of the event.

Regardless of the type of competition a breeder (*casteador*) prefers to enter, he must breed and raise the fighting birds he would like to take to the cockpit. From the time chicks are born, breeders pay attention to their behaviour and identify them individually. This allows them to form a tentative view of how each rooster might develop. The first stage is one of pure nurture but, sooner or later, the cohabitation of chicks (*pollos*) becomes untenable, and they must be kept individually.

Breeders continue to care for their fighting birds, but the birds will be required to prove themselves as fighters through a series of interventions. This includes sparrings (*pechas*), where breeders try to determine whether the stags have the quality to fight in a competition. Birds need to spar more than once, on different days, and with different adversaries. It is of critical importance to see how gamecocks react to other gamecocks and their willingness to keep fighting under different and difficult conditions. Although breeders have favourite characteristics (e.g. looks, proportions), this is a matter of aesthetics. They are more interested in the way that a gamecock fights irrespective of whether it is a fine physical specimen or not. The quality of continuing to fight whatever its injuries is greatly admired and appreciated by breeders and other aficionados. This is an essential aspect, and revelation, of their *casta*. *Casta* is a complex term that refers to bloodline but also to a complex amalgam of animal qualities, in particular their willingness to keep on fighting under extremely harsh conditions.

6.4.5 Chicken and Religious Rituals

Another type of performance involving animals, and especially chickens, is religious ritual. In the Afro-Brazilian Candomblé religion, chickens are relevant in the cosmology and mythology. As preferred sacrificial animals, they are present in a variety of ritual occasions. In a popular myth, the earth is said to have been created with the help of a chicken who spread and clawed the soil over the oceans. Another myth recounts how a guinea fowl, with its spotted feathers and strange appearance, managed to scare off Death itself. Through these cosmological views, chickens and guinea fowls became religious symbols of nourishment and life. The Afro-Brazilian Candomblé draws many of its elements from traditional African religions, especially those widespread in Nigeria, Benin, Angola and Democratic Republic of Congo. Candomblé practitioners worship deities (*orishás*) connected to natural elements like the ocean, the forest, the river and so on, and full membership is possible only after an initiation ritual. In many religious and ritual contexts, the initiation is considered a form of social and spiritual rebirth.

During this ritual, the novice's body is painted with white spots, and a special conic artefact made of herbs and other ingredients (*adoxu*) is placed on the top of the novice's shaved head. These symbols recall the helmet and the feathers of the guinea fowl. The social and symbolic death, which is implicit in the initiation process, culminates in a ritual where the novice is said to be reborn as a guinea fowl, while an actual bird is sacrificed to propitiate the birth of a new life (Vogel et al., 1993).

Yearly Candomblé celebrations honour deities with food offerings and animal sacrifices, which serve to nourish both the deities and the religious community. The chickens are chosen according to the colour of their feathers, their sex and age, and vary depending on the deity who needs to be worshipped. For example, Oxalá (*Oshala*), the deity of purity and silence, will only accept chickens with white feathers and food like white maize and yam. Oxum (*Oshum*), the goddess of love and fertility, will only accept hens, and one of her favourite dishes is composed of black-eyed peas and eggs. Sacrificed chickens also need to be prepared in a specific way. After being plucked, the vital organs and extremities of the bird (e.g. head, feet, wings and tail) are set aside, cooked in a specific way and offered to the deities, as these body parts are considered full of life energy. While preparing the offerings, practitioners ask for health and prosperity for themselves, making the connection between the chicken and the human body (Capponi, 2020). In tracing these multiple correspondences between fowls, humans and deities during rituals, practitioners develop a complex relationship with these birds, which they consider a source of nourishment, a powerful symbol and a way to connect to the invisible world at the same time.

6.4.6 Conclusion

Around the world, chickens are valued for more than their meat or their productive characteristics. Depending on the different cultural understandings, humans appreciate these birds for features such as the colour and beauty of their feathers, their tameness or fighting skills, their behaviour, and symbolic traits. These ideas do not exclude their role as a source of food. Indeed, hobby farmers enjoy the eggs they collect in their back garden; chickens help Amhara women achieve economic empowerment; and Candomblé practitioners use the sacrificial meat to feed the whole community. However, these actions need to be understood as complementary to a whole set of interactions and meanings associated with animals.

6.5 Conclusion

There is more to the roles of other species than friend or food. They work alongside humans, improve access to education, support gender inclusivity and reduce poverty and food insecurity. They may be treated with care or left to fend for themselves. One cannot simply reduce the world of HAIs to good/bad dichotomies, because the world is much more nuanced. Gamecocks may be as beloved as one's pet dog, depending upon the context (Herzog, 2011). A cross-cultural lens allows us to appreciate these subtle differences.

To avoid ethnocentric interventions, any research in this field must work with local communities and individual owners to identify their values and appreciate underlying challenges to change. Empowering animal-owning communities to reflect on their situation and develop their own actions can mitigate ethical concerns. Doing so can also improve outcomes by uncovering solutions that do not conflict directly with local norms. Recognising that interventions have the potential to interfere with community members' lives, individuals working in HAIs must respect people's right to choose by supporting personal agency.

Additional Resources

Link to 360-degree virtual reality video in a brick kiln www.youtube.com/watch?v=BtLdTKPvJi0&t=7s

Invisible Helpers, Working livestock in sustainable development www.youtube.com/watch?v=n0eCORdtfMY&t=82s

Compassionate Handling Factsheets

- Horses, donkeys, and mules www.thebrooke.org/sites/default/files/CompassionateEquineHandlingWeb.pdf
- Livestock www.thebrooke.org/sites/default/files/CompassionateLStockHandlingWeb.pdf
- Pets www.thebrooke.org/sites/default/files/CompassionatePetHandlingWeb.pdf

OneHealthInitiative.com
Onewelfareworld.org

6.6 Learning Outcomes

Source: Illustrated by Loïc Maréchal adapted from istockphotos.

Chapter Summary

- Current bias influences knowledge around human–animal interactions.
- A cross-cultural perspective on human–animal interactions is important to comprehensive understanding.

- Dogs, equids and chickens can act as pets, as a tool for gender empowerment, as a ritual symbol, or as sports animals depending on the context.
- Within cultures, there remain large individual variations in human–animal interactions.
- A One Health/One Welfare approach can help improve outcomes for humans and other animals.

Check Your Understanding

Multiple Choice Questions

Question 1: Which statement is true of most dogs in the world?
  - Answer a: They are kept as pampered pets in loving homes.
  - Answer b: They are neutered and vaccinated to prevent disease.
  - Answer c: They live in free-ranging communities with little to no care from humans.

Question 2: What do working equids contribute to human livelihoods?
  - Answer a: Sustainable transportation and leisure activity.
  - Answer b: Pet keeping and religious rituals.
  - Answer c: Food security and gender inclusivity.

Question 3: Women in the Amhara regions prefer to keep chicken breeds that:
  - Answer a: are good for scaling up production.
  - Answer b: have suitable traits for sacrifice and consumption.
  - Answer c: require lots of care.

Open Questions

1. How does a cross-cultural perspective enhance our understanding of HAIs? Why is this important to the field?
2. Provide a critical analysis on how some animal species are portrayed in different cultures, and how important it is to integrate cross-cultural variations in HAI for animal welfare and/or conservation efforts.

7 Criminal Issues in Human–Animal Interaction

Laëtitia Maréchal, Ross M. Bartels, Georgina Gous and Angus Nurse

7.1 Introduction

Although animal sentience recognition, referring to animals' ability to perceive, feel and experience (Broom, 2014), helps us better understand and improve human–animal interactions (HAIs), most legislation (i.e. laws) still considers animals as inferior living forms or objects (Probyn-Rapsey, 2018). However, changes are slowly occurring at international and national levels as better legislative protection regarding animals is gradually being implemented. This is often the result of extensive research and ethical debates around animal welfare and conservation, as well as the developing of a better understanding of how our treatment of animals is linked with other criminal activities (e.g. animal abuse and domestic abuse; see Section 7.3 in this chapter). This chapter provides an overview of HAIs within a forensic context, ranging from crimes against animals, such as wildlife crime and animal abuse, to the use of animals in criminal investigations.

7.2 Wildlife Crime

7.2.1 Definitions

Wildlife crime can be broadly defined as the illegal exploitation of wildlife species, including poaching (i.e. illegal hunting, fishing, killing or capturing), as well as abuse and/or trafficking of wild animal species. In this chapter, we define wildlife as any non-domesticated non-human animals (Décory, 2019; see also Chapter 11, this volume). However, it should be noted that legislative definitions of wildlife vary across jurisdictions and in academic discourse. For example, some definitions exclude fish, and other definitions define wildlife as including fauna and flora (see discussion below of CITES).

Thus, for an act to be considered a wildlife crime, it must include the following elements (Nurse & Wyatt, 2020):

1. Something that is proscribed by legislation.
2. An act committed against or involving wildlife (e.g. wild birds, reptiles, fish, mammals, plants or trees) which form part of a country's natural environment or which are of a species that are visitors to the country (i.e. non-native species), but that do live in a wild state.
3. Involve an offender (individual, corporate or state) who commits an unlawful act or is otherwise in breach of obligations towards wildlife.

DOI: 10.4324/9781003221753-7

These elements clarify that wildlife crime is considered a social construction because it relates to violation of existing laws created by a country. Therefore, the definition of wildlife crime can vary from country to country. The definition of a crime can also change anytime laws are modified due to, for example, contemporary conceptions and changing social or political conditions. As such, biological principles about animal welfare are largely absent in wildlife crime laws, which has raised criticisms about political and trade interests dominating over wildlife interests. For example, the United Kingdom historically allowed fox hunting with dogs, but this practice was banned in 2004 (Hunting with Dogs Act, 2004). However, this Act could simply be repealed by the government and hunting with dogs could become legalised again depending on the political interests. The sociolegal classification of crime as defined by the criminal law (Situ & Emmons, 2000) also means that any behaviour not prohibited by law is not a crime. Thus, for example, the killing of wildlife within regulated hunting activities (e.g. trophy hunting) or 'pest' control might not constitute a crime as long as the regulatory provisions are complied with (e.g. not using any prohibited methods of taking wildlife, complying with humane killing methods). In this context, wildlife crime has clearly defined notions of victimisation in respect of the non-human animals that may be killed, taken or otherwise exploited, and those which may not. Therefore, laws specify which non-human animals are classified as 'victims' in respect of being targets of prohibited acts and which animals are fully protected.

7.2.2 International Laws and National Laws

International law defines the obligations of states to meet and apply certain legal standards, mostly through treaties and conventions, as the main 'hard law' instruments. Therefore, to establish the core requirements for legal protection of wildlife in international law, different steps must be achieved including examination of any relevant legal documents (e.g. guidance notes, policy documents and any decisions of judicial or quasi-judicial mechanisms). General animal protection law is absent from international law discourse, but a range of wildlife laws exists, operating as three themes (i.e. conservation, criminal and trade laws) where states have agreed a need for international law measures to provide for sustainable use of wildlife. For example, the European Union's (EU's) Habitats Directive (Council Directive 92/43/EEC) contains measures on the conservation of wild fauna and flora intended to maintain biodiversity. The EU's Directive 2008/99/EC on the protection of the environment through criminal law also sets out a strengthening of criminal penalties for environmental crimes. Other international law mechanisms exist such as the GATT/WTO wildlife Trade Agreements, the International Convention on Whaling, and the Convention on the Conservation of Migratory Species, which also provide some wildlife protection.

When discussing wildlife crime, the Convention on International Trade in Endangered Species of wild fauna and flora (CITES) is widely considered the main international law that governs the trade in wildlife and is identified as a key mechanism for dealing with wildlife trafficking (Wyatt, 2013). Brought into force in 1975, CITES tracks the amount of trade from member countries and seeks to monitor illegal activity in respect of wildlife trade (Figure 7.1). CITES provides the international law framework for regional and national legislative wildlife trade protection and also provides the context in which national law enforcement activities on wildlife trade priorities operate. Therefore, the function of trade law is arguably to define legal notions of harm and criminality but not to provide for the health and welfare of animals. In this regard, the logic of international wildlife law 'is not simply to protect endangered species because they are endangered; it is to manage these "natural resources" for human use in the most

Figure 7.1 QR code link to what is CITES? (www.youtube.com/watch?v=zWRA9N-pB9Y), you can also visit the CITES website (https://cites.org/eng).

equitable and least damaging manner' (White & Heckenberg, 2014, p. 133). The principle of 'sustainable use' permeates wildlife law even at the international level; thus, the aim of international wildlife law aim is often to regulate use of wildlife rather than to criminalise it. This is arguably explicit in CITES, which regulates the trade in endangered species of wildlife and allows for continued exploitation of wildlife subject to certain restrictions and compliance with the provisions of CITES.

The Convention has three degrees of protection:

1 Prohibits international commercial trade of species listed in Appendix I of CITES (higher level of protection).
2 Allows commercial trade of species listed in Appendix II, subject to export or re-export permits.
3 Allows commercial trade of species listed in Appendix III, subject to export or re-export permits, or certificates of origin.

As of 26 November 2019, there were 5,945 non-human animals listed in the CITES appendices. There were 687 in Appendix I, 5056 in Appendix II, and 202 in Appendix III (CITES, 2021). CITES is then implemented in different legislation for each country. For example, CITES is implemented in the EU through a set of regulations known as the EU Wildlife Trade Regulations (EC No 333/97; EC No. 865/2006 and EU No. 791/2012). The regulations therefore provide for a CITES enforcement regime, creating powers for the police to take action against illegal trade in wildlife in countries adhering to the EU. The incorporation of international law mechanisms such as the CITES provisions into national legislation also means that, at least in principle, wildlife is protected from only certain harmful activity (e.g. the illegal trade in wildlife), and illegal activity affecting wildlife is also subject to criminal sanctions.

One difficulty with legislation related to wildlife is its intended use as conservation or wildlife management legislation rather than as species protection and/or criminal justice legislation. Sollund (2013, p. 73) argues that 'CITES can be criticized for legitimating trade and trafficking in animals and for prolonging and encouraging abuse and species decline by regarding non-human species as exploitable resources'. National wildlife law, while implementing international and domestic perspectives on wildlife protection, routinely allows for the continued exploitation of wildlife. For example, while national laws provide open seasons to take selected wildlife, or prohibits some methods of killing/taking wildlife such as traps, they also allow other methods to be used, such as shooting (Nurse, 2012). Therefore, the extent to which laws are enforced is as much a political decision as a moral one, based on acceptance of the notion that humans owe a duty towards other inhabitants of the planet (Benton, 1998). Animals' status as property still dominates their legal protection such that animal protection and anti-cruelty laws 'take the people who own or use animals as primary objects of moral concern, rather than the animals themselves' (Rollin, 2006, p. 155–156). Therefore, animal laws, whether conservation, wildlife management or wildlife protection law, generally provide protection for animals largely commensurate with human interests (Nurse, 2013; Radford, 2001; Schaffner, 2011) and are socially constructed to achieve varied ends (Nurse, 2011).

7.2.3 Wildlife Crime: Offending Types and Behaviours

Wildlife crime may be considered to have a single or primary cause, namely exploitation of wildlife for profit by offenders who make a rational choice to commit a crime. However, green criminology, which studies environmental harms and crimes affecting the environment and non-human nature, has identified that 'there is much to be learned from case studies of poaching, the illegal trade in wildlife and the impacts of these activities' (Lynch et al., 2015, p. 1187). This includes consideration of the range of offending types and behaviours that exist (Nurse, 2011, 2013; Wyatt et al., 2020). While some offences incorporate the taking and exploitation of wildlife for profit (i.e. wildlife trade), others involving the killing or taking or trapping of wildlife either in connection with employment (e.g. bird of prey persecution, ranching and livestock production) (Roberts et al., 2001, p.27) or for purposes linked to field sports (e.g. hunting with dogs, large carnivore hunting, shooting). Five main categories of wildlife offender have previously been defined (Nurse, 2011; 2013, pp. 69–70):

1. Traditional profit driven offenders – those who are involved for direct benefit.
2. Economic criminals – those whose offending is linked to economic considerations such as a need to break laws in order to retain employment.
3. Masculinities criminals – those whose offending is directly linked to their masculinity and perceptions of 'appropriate' male behaviours.
4. Hobby offenders – those who derive limited financial benefit, but for whom the activity is a hobby or compulsion.
5. Stress offenders – those who commit offences as a result of their own victimisation.

Evidence shows that motivations vary between different types of offenders, and between individual and organised/group offenders. For organised crime groups, wildlife trafficking is primarily a profit-driven offence, undertaken to gain maximum revenue often without the punitive consequences and overt law enforcement attention that accompanies other illegal trafficking activities such as drugs, weapons or human beings (Wyatt et al., 2020). These crimes also fit the masculinities offender model, where crimes are committed to reinforce notions of power and masculinity and are rarely committed by lone individuals (Nurse, 2013, p. 75).

In addition to the different motivations driving offenders, they also have varied justifications for the crimes they commit (Table 7.1). Sykes and Matza's neutralisation theory (1957) provides for contextual understanding of the justifications used by offenders that gives them the freedom to act (and a post-act rationalisation for doing so). This means that offenders neutralise the harm of their actions by relying on different justifications or explanations that minimise their actions,

Table 7.1 Outline of the various motivating factors for wildlife crime by offender type arising from empirical research and literature analysis

Type of Criminal	*Ignorance of the Law*	*Pressure from Employer or Commercial Environment*	*Financial Gain*	*A Feeling of Power*	*Excitement Thrills or Enjoyment*	*Low-Risk Crime*	*Keeping Tradition or Hobby Alive*
Traditional Criminal	No	No	Yes*	No	Yes	Yes	No
Economic Criminal	No	Yes*	Yes	No	No	Yes	Yes
Masculinity Criminal	No	No	No	Yes*	Yes	Yes	Yes
Hobby Criminal	No	No	No	No	Yes	Yes	Yes*
Stress Offender	No	No	No	Yes*	Yes	Yes	No

(* indicates the primary motivator).
Source: Adapted from Nurse, 2011, p. 48.

for example, by claiming that harm caused to wildlife is a victimless crime, or by arguing that their offending is not a real crime and enforcers should direct their attention elsewhere. Offenders may also argue that their activities are necessary in order to preserve a tradition or way of life. Indeed, in some contexts, such as traditional hunting communities, the killing of wildlife can be considered a legitimate form of resistance (von Essen & Allen, 2017) (e.g. dog fighters, cockfighters and hunting communities: Eliason, 2003; Forsyth & Evans, 1998; Hawley, 1993). In some cases, offending can be a result of learned behaviour. Sutherland's (1973) differential association theory suggests that those involved in organised groups learn their activities from others in their community or social group (Sutherland, 1973). Such learning can influence the continuation of activities that are considered to be part of a community, a way of life or which are linked to other activities. Indeed, organised crime culture itself may reinforce the view that there is no harm in continuing with an activity that simply represents another form of import–export activity, utilising existing trade routes. Wildlife trafficking can arguably be rationalised as less harmful or serious than human trafficking or drug smuggling. Awareness of the illegal nature of the offender's actions leads to the justifications outlined by Sykes and Matza (1957). However, the association with other offenders, the economic (and subcultural) pressures to commit offences, and the personal consequences associated with failures are strong motivations for organised crime participants to commit wildlife trafficking offences (Merton, 1968).

However, it should be noted that organised crime involved in wildlife crime is also not a single homogenised group. South and Wyatt's (2011) comparative analysis of criminal actors in the illegal drug trade and the wildlife trade, identified several key actors defined as: trading charities, mutual societies, business sideliners, criminal diversifiers and opportunistic irregulars (South & Wyatt, 2011). More recently, Wyatt et al. (2020) identified three types of organised crime groups: organised, corporate and disorganised groups. These groups are arguably involved in wildlife trafficking at different levels and in different ways. For example, some groups are takers or procurers of wildlife, some are suppliers, and some are distributors or transporters. Wyatt et al. (2020, p. 353) note that the nature of the organisation and of wildlife crime are often not defined as 'serious' according to the UN definition of seriousness as 'conduct constituting an offence punishable by a maximum deprivation of liberty of at least four years or a more serious penalty' (United Nations Office on Drugs and Crime [UNODC] 2004, p. 5). As a result, enforcement activity needs to consider the varied nature of offending and organised crime to implement a response aimed at both the 'serious' and the less 'serious' crimes.

Analysis of wildlife law and wildlife crime policy responses identifies some possible shortfalls in the use of laws. Therefore, criminality might not be the best tool to address all wildlife crime issues. As indicated, wildlife law prohibits only certain acts whilst allowing for continued exploitation of wildlife. Wildlife laws also broadly adopt a punitive approach consisting of fines and prison sentences rather than being aimed at addressing the harm to wildlife and non-human nature. Green criminological research has begun to explore the efficacy of such an approach and to consider whether alternative approaches may be more effective. This includes approaches focusing on the harm caused to the environment and wildlife, and seeking to impose sanctions aimed at repairing that harm.

7.3 Animal Abuse

Animal abuse can be defined as 'all socially unacceptable behaviour that intentionally causes unnecessary pain, suffering or distress and/or death to an animal' (Ascione, 1993, p. 83). Companion animals are thought to be the most frequently abused (van Wijk et al., 2018), with hitting, kicking and/or shooting being the most common forms of abuse (Hensley & Tallichet, 2005). Unfortunately, due to being under-reported and under-investigated, accurate rates of animal abuse are not well established (van Wijk et al., 2018). Conviction rates also tend to be low. For example, in the United Kingdom, around 1.5% of investigated cases ended in a criminal conviction (Royal Society for the Prevention of Cruelty to Animals, 2015). This is because it is difficult to accumulate prosecutable evidence, as animals are 'voiceless victims' (Alleyne & Parfitt, 2019) and the cause of animal injuries is not easy to ascertain (Intarapanich et al., 2016). Also, not all countries recognise animal abuse in the same way (or even at all), and so not all countries have laws against it. This means that most research conducted on animal abuse is likely biased towards countries that recognise animal abuse as a criminal offence. Furthermore, very limited research on the topic has been published outside the United States, the United Kingdom or Australia (Pręgowski & Cieślik, 2020).

7.3.1 Animal Abuse and Intimate Partner Violence

Much of the research on adult-perpetrated animal abuse focuses on its relationship with intimate partner violence (IPV) and tends to be based on accounts provided by IPV victims (Faver & Strand, 2003a; Febres et al., 2014). These studies suggest that the rate of co-occurrence between animal abuse and IPV ranges from 25% (Simmons & Lehmann, 2007) to 86% (Strand & Faver, 2006), which has led to the view that animal abuse is a 'red flag' for other forms of domestic violence (DeGue & DiLillo, 2009). This observation has been explained using the 'violence graduation hypothesis' (Ascione, 2001), which states that animal abuse acts as a 'stepping stone' to violence against humans and, therefore, must precede violence against humans.

Based on this, animal abuse would be assumed to occur before IPV. However, the abuse of animals in this context is typically used to coercively control, punish, threaten and/or emotionally abuse a partner (Faver & Strand, 2003b; Fitzgerald et al., 2022). Thus, rather than preceding

Figure 7.2 QR code video to illustrate the link between animal and domestic abuses in Australia (The link between animal abuse and domestic violence). Warning: Sensitive topic discussed; viewer discretion is advised!

IPV, animal abuse may be a part (or form) of IPV. Furthermore, Febres et al. (2014) found that animal abuse did not predict IPV perpetration over and above antisocial traits and alcohol use. These findings do not support the violence graduation hypothesis. Indeed, other researchers have argued that the 'deviance generalisation hypothesis' is a better explanation for the link between animal abuse and other forms of violence (Figure 7.2).

The deviance generalisation hypothesis states that 'animal abuse is one of many antisocial behaviours committed by an individual' (Arluke et al., 1999, p. 970). Importantly, this hypothesis is supported by research. For example, animal abuse has been found to be associated with other offending behaviours, such as property and drug offences (Arluke et al., 1999; Vaughn et al., 2009). A meta-analysis also found that animal abuse was associated with both violent and non-violent offending (Walters, 2013). More recently, van Wijk et al. (2018) found that half of their sample of animal abuse perpetrators had also committed other offences (e.g. property, vandalism and traffic offences). Counter to the 'graduation hypothesis', the abuse of animals tended to occur after the commission of these other offences.

7.3.2 Understanding Animal Abuse

The other half of van Wijk et al.'s (2018) sample were first-time offenders (i.e. they had not been registered for any other offences). Therefore, to gain a fuller understanding of animal abuse, it is important to establish the motivations and facilitatory factors of animal abuse. Ford et al. (2021) investigated the offence pathways associated with animal abuse in a community sample. They found that, on the day of the incident, participants reported being intoxicated, distressed, bored or preoccupied. This made them susceptible to perceiving an animal's behaviour as hostile, threatening or disobedient, leading to negative feelings (e.g. anger). As a result, a decision to abuse the animal was made in order to dominate, punish or express negative emotions towards the animal. The abuse tended to be physical (e.g. hitting, pushing, dragging, throwing and/or killing the animal), and was accompanied by feelings of anger, fear, guilt or amusement. These findings provide a clearer account of why and how animal abuse unfolds, including associated cognitions and emotions.

Other studies have examined the cognitive, affective and personality factors associated with animal abuse. Regarding cognition, animal abuse tends to be associated with the belief that animals are there to be mastered (Raupp, 1999) and that aggression is normal (Sanders & Henry, 2018), as well as positive attitudes towards the mistreatment of animals (Alleyne et al., 2015).

Regarding affective factors, Parfitt and Alleyne (2018) found that anger regulation and impulsivity are associated with a propensity to abuse animals. Low self-esteem and empathy levels are also associated with animal abuse (Alleyne et al., 2015; Alleyne & Parfitt, 2018). Low empathy is inherently linked to callous-unemotional traits and psychopathy, thereby linking personality with animal abuse. Indeed, callousness is associated with animal abuse in men (Gupta, 2008) and the propensity to abuse animals in women (Ireland et al., 2022). This supports other studies showing that, in both men and women, psychopathy is related to animal abuse in community (Kavanagh et al., 2013) and incarcerated samples (Rock et al., 2021). These findings provide the groundwork for future research and the generation of a multifactorial theory. They also have important implications for policy, risk assessment and forensic practice.

7.3.3 Sexual Abuse of Animals

Sexual interactions between a human and an animal can also be regarded as animal abuse (van Wijk et al., 2018). Indeed, Stern and Smith-Blackmore (2016) stated that 'animal sexual abuse' involves harm inflicted on an animal for the purpose of human sexual gratification. Other terms include bestiality and zooerasty. Sometimes sexual contact with an animal is referred to as zoophilia, although this is inaccurate. Zoophilia is a distinct concept defined as an intense and persistent sexual interest in animals (American Psychiatric Association, 2013). Although incidences of human–animal sexual contact are associated with a sexual interest in animals (Seto et al., 2021), not everyone who engages in this behaviour has zoophilia (Holoyda et al., 2018). Other reasons for engaging in sexual contact with animals include curiosity, expressing love or affection, and wanting affection (Miletski, 2002). Those who wish to express their love for an animal would, arguably, not view the sexual contact as abusive because it does not necessarily result in pain, injury or death for the animal. Indeed, many people who identify as having zoophilia refer to 'bestiality' as an abusive sexual act, as it is devoid of an emotional bond (Holoyda & Newman, 2014). Nevertheless, sexual contact with an animal is widely viewed as abusive because it can cause injury and even death (Stern & Smith-Blackmore, 2016) and because it is not possible to gain consent from animals (Beirne, 2000).

The prevalence of human–animal sexual contact appears to be low within the general population (Bouchard et al., 2017), ranging from 2% to 3% for women and 5% to 8% for men (Szymańska-Pytlińska et al., 2021). The types of animals that people have sexual contact with range from companion animals (e.g. cats, dogs) and farm animals (e.g. sheep, pigs, horses), to more exotic animals (e.g. dolphins) – although dogs and horses tend to be more common (Williams & Weinberg, 2003). In forensic populations, the rates of human–animal sexual contact are higher. For example, English et al. (2003) reported a rate of 36% in a sample of people convicted of a sexual offence. Seto and Lalumiere (2010) reported a 14% rate of 'bestiality' among adolescents who have sexually offended. These findings suggest that human–animal sexual contact (particularly in childhood) may be a risk factor for sexually offending against humans in adulthood. However, Hensley et al. (2010) found that those who engaged in 'childhood bestiality' were more likely to commit a range of non-sexual interpersonal offences against humans in adulthood (e.g. robbery, assault, murder). Thus, as with non-sexual animal abuse, human–animal sexual contact could be explained by the generalised deviance hypothesis. However, as noted by Beetz (2005b), the motivations behind human–animal sexual interactions are varied and are not always based on a desire to harm an animal or gain sexual pleasure. Some people who engage in this behaviour even believe that animals have an intrinsic ability to give their consent to sexual activity (Sendler, 2018). This indicates that there may be subtypes of people who engage in sexual contact with animals. Since human–animal sexual contact is an animal welfare issue, it clearly needs to be better understood from an academic standpoint. Thus, before any solid conclusions can be formed, more research is needed.

7.4 Use of Animals in Forensic Settings

This section will address the use of animals in forensic settings. It will begin by discussing the use of detection dogs by law enforcement professionals in forensic investigations before moving on to evaluate the effectiveness of this technique. Then the use of animal evidence to understand criminalist issues will be explored. Specifically, we will address forensic entomology, animal experimentation and animal DNA.

7.4.1 Detection Dogs

Dogs have been used for domestic purposes for thousands of years (see Chapter 9, this volume), but more recently they have been used by law enforcement professionals in forensic investigations (Singh & Kaur, 2015). Dogs have a highly developed sense of smell, making them particularly ideal for use in forensic investigations (Singh & Kaur, 2015). Detection dogs, also known as sniffer dogs or K9s, are dogs that are trained to detect and locate articles of interest, ranging from illegal drugs to missing people, using their sense of smell (Furton & Myers, 2001). Detection dogs are used in most countries around the world. Dog breeds mostly used by law enforcement include, for example, Belgian Malinois dog, bloodhound, Border Collie, Doberman Pinscher, German Shepherd, Golden Retriever, Labrador Retriever, and the English Springer Spaniel (Allsopp, 2012).

7.4.1.1 Training Detection Dogs

Training detection dogs is a lengthy process and begins with training for the dog handler to ensure that they train the dog to the best of its ability (Lit et al., 2011). A suitable dog also needs to be selected based on physical (e.g. free from health conditions), temperamental and behavioural standards (e.g. evidence of inquisitiveness, highly responsive, friendly to people) (Jezierski et al., 2014). In addition, a positive relationship between the dog and the handler is crucial for the success of the training. Detection dogs are trained predominantly using associative learning techniques, linking two previously unrelated stimuli, either through classical or operant conditioning (McGreevy & Boakes, 2011). In a forensic context, classical conditioning training involves repeatedly presenting a target odour to the dog whilst teaching them to display a particular behaviour (e.g. sitting) upon detection of that odour. Operant conditioning is typically established through providing the dog with rewards (e.g. food or playing with the dog), for the correct response, (e.g. sitting), upon successful detection of the target odour (Hayes et al., 2018). Withholding a reward, a form of negative punishment, when the dog elicits a weak response is also used to avoid inadvertently training the dog to perform false positives, such as sitting in the absence of an odour (Haverbeke et ál., 2008).

7.4.1.2 Specialised Detection Dogs

Detection dogs are trained to search for many things depending on their specific role in the forensic investigation (Table 7.2).

Table 7.2 Description of the different specialised roles of detection dogs

Role	*Description*
Drug Detection Dog	Trained to detect the presence of illicit substances and can locate even the tiniest traces of a drug.
Explosives Detection Dogs	Used more in recent years due to the increased attention to terrorism, dogs are trained to detect odours of specific substances such as sulphur, nitroglycerine, and any other compound used in the production of gunpowder and explosive devices.
Arson Dogs	Trained to detect chemical traces of accelerants and locate its source.
Cadaver Dogs	Trained to follow the scent of decomposing flesh to locate the bodies of deceased humans.
Search and Rescue Dogs	Used to locate living individuals, usually missing people and those trapped during mass disasters.
Tracking Dogs	Used to track down living individuals such as suspected criminals.
Wildlife Detection Dogs	Used to locate specific animals or plants, or trace such as faeces, nests. These dogs can be used in conservation or wildlife crime settings.

7.4.1.3 Effectiveness of Detection Dogs in Forensic Settings

Compared to advanced analytical chemical detection, detection dogs are considered the most effective means of finding articles of interest due to their excellent search ability. For example, research has shown that detection dogs are highly effective at discovering explosives and other contraband in places such as airports (Helton, 2009; Oxley & Waggoner, 2011). Furthermore, Jezierski et al. (2014) showed that drug samples can be indicated by dogs after only 64 seconds of searching time, with 87.7% of indications being correct and only 5.3% being incorrect (i.e. false positive). Successful recovery rates of scattered human remains have also been shown to range between 57% and 100% (Komar, 1999). Detection dogs are also far less costly compared to machinery that would otherwise be used to detect articles of interest (Hayes et al., 2018).

However, the detection abilities of dogs can be negatively influenced by several factors. For example, dogs may be unable to work in some temperature and humidity conditions, or they may only be able to work for very limited periods of time before they need to rest (Hurt & Smith, 2009). Dogs are also vulnerable to distractions in the environment such as loud noises, bright lights and new surroundings (Hayes et al., 2018). Furthermore, handler error may mislead the dog into false identification, and the probability to commit such an error can be increased by the stress of the handler (Zubedat et al., 2014).

7.4.2 Animal Evidence

Evidence can be defined as anything that can prove, or disprove, a fact or contention (Gardner, 2011). In forensic settings, evidence can be used to identify victims and suspects, to prove the guilt or innocence of a suspect (Touroo & Fitch, 2016), and to understand criminalist issues. Many different pieces of evidence can be obtained from a crime scene, including taking photographs and physical measurements of the scene, and collecting forensic evidence such as DNA. Investigators may also decide to use a forensic entomologist or animal experimentation to help them to understand criminalist issues.

7.4.2.1 Entomology

Forensic entomology is the study of the application of insects and other arthropods in criminal investigations (Catts & Goff, 1992). In most of these investigations, a forensic entomologist becomes involved with the legal system after the initial discovery of a body or other crime (Haskell, 2007). This is because, in many cases, insects can be used to determine a time of death. Research suggests that insect larvae present on a dead body can provide evidence for the estimation time interval between death and corpse discovery (also called post-mortem index) for up to one month (Amendt et al., 2004). Insect data can also be used for determining the site of the crime due to differences in the species of insects involved with a decomposing corpse in different habitats and environments (Haskell, 2007). If a species associated with one type of habitat is present on a corpse but is found to be different from those that are part of the habitat where the body is found, this could indicate that the corpse was transported after the victim's death. A third means by which a forensic entomologist becomes involved in a case is by being hired as an expert witness by the prosecution or the defence after charges have been filed (Haskell, 2007). Such information provided in court may be particularly relevant in determining the guilt or innocence of a defendant.

7.4.2.2 Animal Experimentation

More than a hundred million animals are used for testing in laboratories around the world each year, largely fuelled by the belief that using animals in experimentation leads to results that are safe, sound and useful for humans (Cattaneo et al., 2015). Animal experimentation is present in nearly every field of science, and whilst the significance of findings from clinical research may, in some instances, outweigh the sacrifice and suffering of animals, the use of animals for experimentation still has serious problems with validity and practical application (Cattaneo et al., 2015). Within the forensic field, animals have been used predominantly to understand crime-related issues. Earlier examples of this include using animals to inflict wounds and burns for dating lesions and drowning unanaesthetised dogs (see Knight, 1992). In a more recent review, however, there was still evidence of many ethically questionable cases where animals have been used in forensic experimentation. Examples include the administration of pesticides (Yoshimoto et al., 2012), ante mortem production of traumatic brain injury (Li et al., 2013), blunt injury to the precordial regions in dogs (Guan et al., 2007), air embolism in rabbits (Xiang et al., 2013), gunshot or stab wounds to pigs (Prat et al., 2012), hanging (Matsumoto et al., 2013), electrocution (Huang et al., 2012), asphyxia (Nagai et al., 2011), drowning (Hayashi et al., 2009), and hypovolemic shock (Kenji et al., 2012) in mice rabbits and dogs.

Therefore, some measures have now been implemented in some countries to minimise animal suffering in research. For example in the United Kingdom, the Home Office is responsible for legislation in the field of animal welfare. The Animals (Scientific Procedures) Act (1986) requires experimenters to consistently follow the 3 Rs (i.e. reductions, refinement, replacement: see Chapter 5, this volume). Researchers are required to demonstrate that consideration has been given to replacing animals where possible, reducing the number of animals, and refining methodology to minimise suffering. Although there are some scientists who advocate for the use of animals in forensic research, some have called for the use of animals to face even more scrutiny and, in some instances, be stopped altogether.

7.4.2.3 Animal DNA

DNA, or deoxyribonucleic acid, is the molecule that contains the unique genetic code of any live organism, including animals (unless identical twins), plants (unless they are able to propagate by asexual means), protists, archaea and bacteria. Once a DNA profile has been extracted from the sample evidence, it is checked against a database of other profiles. If there is a match, then it could be used as evidence in a criminal case and in a court of law.

Human DNA evidence has been used to successfully solve many criminal cases since 1986. Animal DNA evidence is also now becoming more commonplace as investigators realise that the same techniques used in human DNA analysis can be applied to animal evidence too. The first time in forensic history that animal DNA was used to identify a criminal was in 1994 by the Royal Canadian Mounted Police (RCMP) detectives who successfully tested hairs from a cat to link a man to the murder of his wife (Menotti-Raymond et al., 1997).

Animal evidence can help solve a range of cases, each of which has been outlined below:

- **Animal crimes** such as illegal trade, theft, abuse and cruelty from humans (see wildlife crimes and animal abuse sections above).
- **Animal attacks on humans.** Forensic procedures are conducted to establish whether the crime is linked to an animal attack and the type (species or individually identified), or whether the animal markings found on a victim are unrelated to the crime.
- **Human-on-human crimes.** Animal DNA evidence can provide an indication as to where the crime was committed or whether there has been physical contact between the victim and a suspect.

DNA evidence has long been considered the 'gold standard' of forensic evidence due to its ability to exonerate the innocent and convict the guilty (Lynch, 2003). With new highly sensitive technology, even the tiniest amounts of DNA evidence, often from just a few cells, can be collected from crime scenes and used to aid in the efficiency of solving criminal cases (Stiffelman, 2019).

However, there are several factors that can affect the quality of DNA left at a crime scene, including environmental factors such as heat, sunlight, moisture, bacteria, mould and some chemicals (Touroo & Fitch, 2016). Furthermore, although the fundamental aspects of animal forensic DNA testing are reliable and acceptable, animal forensic testing still lacks the standardised testing protocols that human genetic profiling requires (Kanthaswamy, 2015). There is also a lack of consensus about how best to present the results and expert opinion to comply with court standards (Kanthaswamy, 2015).

7.5 Conclusion

In this chapter, criminal issues in HAIs were explored from two angles: animal crimes and the use of animals in forensic settings. Animal crimes are complex, driven mainly by sociopolitical, economic and cultural perspectives, resulting in extreme differences in legislation between countries and animal species. Overall, most animal laws are anthropocentric, and animal needs, in terms of welfare and conservation, are often considered in light of human needs. Therefore, human–animal forensic research has an important role to play to improve animal welfare and conservation, thus, providing evidence to improve animal laws.

The proven reliability and the cost-effectiveness of animal use in forensic settings have led to an increased practice in some countries, either as means to gather evidence (e.g. dog detection) or as evidence themselves (e.g. animal DNA). Although their use might be proven helpful, animal welfare and ethical concerns have been raised, and debates are crucial in advancing our understanding and improving the use of animals in forensic settings.

Additional Resources

Videos on the link between animal abuse and domestic violence: https://vimeo.com/user42222895. Warning: Sensitive topic discussed; viewer discretion is advised!

7.6 Learning Outcomes

Source: Illustrated by Loïc Maréchal adapted from istockphotos.

Chapter Summary

- To be considered as illegal, wildlife crimes and animal abuse must violate existing laws. These laws, and what constitutes an animal crime, vary greatly between countries, and thus animal protection (e.g. animal welfare and conservation), might not always be considered.
- Wildlife crimes commonly include illegal hunting, fishing, killing or capturing, abuse and/or trafficking of non-domesticated animal species.
- Animal abuse is often associated/co-occurrent with other crimes and anti-social behaviours including domestic violence.
- Animals have been proven useful in forensic settings, such as dog detection, forensic entomology and animal DNA, to help solve crimes. However, some ethical debates have been raised around the use of animals in forensic experimentation.

Check Your Understanding

Multiple Choice Questions

Question 1: What does CITES stand for?

- Answer a: Convention on International Trade in Endangered Species.
- Answer b: Conservation on International Traffic in Endangered Species.
- Answer c: Criminology of International Threats in Endangered Species.

Question 2: What is the violence graduation hypothesis?

- Answer a: animal abuse occurs after the commission of these other offences.
- Answer b: animal abuse is one of many antisocial behaviours committed by an individual.
- Answer c: animal abuse acts as a 'stepping-stone' to violence against humans and, therefore, must precede violence against humans.

Question 3: Which animal-related technique is often used to discover hidden drugs or explosives?

- Answer a: DNA analysis.
- Answer b: animal experimentation.
- Answer c: detection dog.

Open Questions

1 Discuss the potential drivers for animal cruelty in the context of either animal abuse or wildlife crimes.
2 Discuss whether animals can be used in forensic investigations, including their advantages and limitations.

8 Animal-Assisted Intervention and Professional Practice

Mirena Dimolareva, Victoria Brelsford and Kerstin Meints

8.1 Introduction and Definitions

This chapter provides an overview of different types of animal-assisted interventions (AAIs) and their definitions, their history and purpose. Having introduced standard definitions of the main fields within animal-assisted activity (AAA), animal-assisted therapy (AAT) and animal-assisted education (AAE), we will highlight benefits as well as potential risks involved in AAI and include discussion of recent results in the field. We will embed our overview and discussion in theories addressing how and why AAI may work. We will furthermore address questions of scientific rigour, and provide recommendations for best practice in AAI, including the safety and welfare of all involved and highlight the need for internationally accepted and recognised standards and guidelines for AAI professionals.

The beneficial value of a dog's presence during therapy sessions with children was first observed by clinical child psychologist Levinson (1969). Since then, academic interest in the benefits of human–animal interaction (HAI) has grown (Beck & Katcher, 1984) and scientific studies have followed (Friedmann et al., 1980) (for an overview, see Fine (2019)). Topic areas have included the role pets can have on children and adults' health and well-being, including socio-emotional and cognitive benefits, and, more recently, the role of pets in older adults' life (Gee & Mueller, 2019) (see also, Chapters 5 and 9, this volume).

AAI in particular have become increasingly popular in recent years and a range of intervention types are used in various settings, with a range of populations, for different purposes. Currently, the field is in the somewhat unusual situation that the speed of application of AAI in practice has overtaken scientific progress – while there is some evidence of beneficial effects, there is also conflicting information and further clarification must be produced. The field of AAI has in the past suffered from a lack of scientific robustness; hence, to assess interventions effectively, and to come to valid conclusions about the true efficacy of AAI, rigorous scientific methods must be applied moving forward.

When discussing AAI, it is important to distinguish between different types of activities involving human–animal interactions (HAIs). The terms animal-assisted interventions (AAI), animal-assisted therapy (AAT), animal-assisted education (AAE), animal-assisted counselling or coaching (AAC) and animal-assisted activities (AAA) have been defined by an International Association of Human-Animal Interaction Organisations (IAHAIO) White paper (2018), as shown in Table 8.1.

The term animal-assisted interventions serves as an umbrella term for interventions that purposely include an animal. Many research studies also fall under this term. Next to general AAIs, such interventions also include the more specific subcategories of AAT, AAC, AAE, and AAA.

DOI: 10.4324/9781003221753-8

Table 8.1 Types of animal-assisted interventions and their definitions and main features

<table>
<tr><td colspan="3">Animal-Assisted Interventions (AAI)
• Umbrella term for planned and goal-oriented and/or structured interventions that include an animal on purpose.
• Many research studies fall under this general term; some employ randomised controlled trials to measure the effects of AAI, while others use before and after measures with their interventions.</td></tr>
<tr><td colspan="3">More Specific Sub-categories:</td></tr>
<tr><td>Animal-Assisted Therapy (AAT)
Animal-Assisted Counselling (AAC)
Animal-Assisted Coaching (AAC)</td><td>Animal-Assisted Education (AAE)</td><td>Animal-Assisted Activities (AAA)</td></tr>
<tr><td>• Involves structured and goal-directed interventions for therapeutic, counselling or coaching purposes.
• Sessions planned in advance
• Aim to achieve specific therapeutic, coaching or counselling goals.
• Integrated into specific therapy (e.g. cognitive behavioural therapy, speech therapy etc.).
• Progress measured to monitor and improve specific outcomes for the recipient.
• Delivered by formally trained and licensed professionals
• Carried out in a variety of settings such as hospitals, clinics, educational settings.</td><td>• Involves structured and goal-directed interventions for therapeutic, counselling or coaching purposes.
• Sessions planned in advance.
• Aim to achieve specific educational goals (e.g. improvement in reading via dog-assisted reading programmes).
• Integrated into teaching day in educational settings.
• Progress measured to monitor and improve specific outcomes for the recipient.
• Delivered by formally trained, qualified education professionals.
• Carried out in a variety of educational settings such as infant, primary, secondary schools, colleges and universities.</td><td>• Involve informal, goal-oriented sessions.
• Sessions booked in advance.
• Delivered by a variety of practitioners who may, or may not, be trained professionals.
• Often involve volunteers, especially where companion animals are registered as visiting therapy animals, mostly dogs.
• Often delivered in a variety of settings such as schools, hospitals, centres for veterans, youth centres and nursing homes.
• May also be involved in crisis response.</td></tr>
</table>

Source: Following Fine, 2015; IAHAIO, 2018.

Importantly, the deliverer of any type of AAI or the animal handler must have some training and sufficient knowledge of their animal partner's behaviour, needs and health (IAHAIO, 2018).

It is worth clarifying that, for instance, a school-based intervention could fall under AAI, AAE, AAT, AAC or AAA depending on the specific type of intervention offered. At times, it is useful to further specify the animal species, for example, when working with horses, the terms *equine*-assisted therapy or *hippo*therapy have been used, or *equine*-assisted activities (see Anderson & Meints, 2016 for further details). When working with dogs, the terms above are often amended to use *dog*-assisted or *canine*-assisted interventions / therapy / education or counselling/coaching. A useful illustration relating to the definitions can be found at Figure 8.1.

The choice of the right types of AAI, consistency and continuity within an intervention are all important for applied practice in therapy and educational settings alike to ensure progress.

Figure 8.1 Video by Molly Tobin discussing the different AAI definitions: www.youtube.com/watch?v=KFIGEC4fb-0.

Consistency and treatment fidelity are also vital in research, with randomised controlled trials representing the gold standard for assessments of intervention efficacy. Importantly, the welfare of all involved, humans and animal partners, has to be considered with great care across all activities.

8.2 Benefits of Animal-Assisted Interventions

We will next provide a spotlight on research in AAI, including subtypes of AAI. However, this overview cannot be exhaustive, but instead serves to highlight recent results in different research areas and serves to whet the reader's appetite for further investigation of the topic.

AAI can involve a range of animals, and those most often employed are dogs, horses and guinea pigs, but other species have also contributed to AAI, such as farm animals, rabbits, donkeys, fish and others (Kapustka & Budzýnska, 2020; Meints et al., 2017). A large range of AAI research involves dogs, due to their popularity as a pet, but also due to the accessibility of volunteers with a pet dog registered as a therapy animal. Whilst including dogs in AAI is common practice, there is often little consistency across studies or applications in terms of the topic of study or specific area of application, or the length, structure, or number of sessions being administered. Nonetheless, beneficial effects are widely reported, albeit not in all studies.

With young children, beneficial effects have been demonstrated in AAI on cognitive tasks such as improved category selection and fewer errors during categorisation of animate versus inanimate stimuli, with improved performance for animate stimuli in the presence of a real dog (Gee et al., 2010; Gee et al., 2012), better compliance to instructions during motor activities (Gee et al., 2009), less need of prompts during memory tasks (Gee et al., 2010) and improved spatial ability (Brelsford et al., 2022). AAI has been used with schoolchildren in various settings and provides mixed results (see Brelsford et al. (2017) and McCune et al. (2017) for reviews).

AAE is typically conducted for reading-age children and teenagers, with educational goals in mind and facilitated by professionals working within education. Again, these can include interaction with a range of animals. Reading with dogs has been carried out as AAE with specific goals in mind but can also function as an activity classed as an AAA as schools often invite a therapy dog with a volunteer handler to sit and read with one child, or a group of children. Research has demonstrated improvement in children's reading ability (Bassette & Taber-Doughty, 2013;

Kirnan et al., 2016; Le Roux, 2014; see also Hall et al., 2016 for a review), citing motivational factors for the child's improved performance.

Changes in physical well-being after AAI have also been widely recorded, for children as well as for adults. For example, physiological changes, including lower levels of cortisol, decreased blood pressure and heart rate, point to a reduction in stress, caused by the presence of a dog (Baker et al., 2010; Beetz et al., 2012; Friedmann et al., 2011; Gee et al., 2014; Meints et al., 2022; Somervill et al., 2009). In older adults, improved mental ability and social interaction and reduction in loneliness, anxiety and depression symptoms could be observed (Banks & Banks, 2002; Friedmann et al., 2011; Gee & Mueller, 2019).

Interestingly, the more direct the contact (i.e. stroking), the lower the levels of cortisol in children and adults (Kertes, et al., 2017; Odendaal & Meintjes, 2003). This opens up the question of whether the length and quality of interaction can lead to differing outcomes.

Children with autism spectrum disorders (ASD) have shown more social interaction through smiles when spending time with a dog (Funahashi et al., 2014), as well as showing reduced anxiety in social situations evidenced by reduced galvanic skin responses when spending time with guinea pigs (O'Haire et al., 2015). O'Haire et al. (2013) also found a stress-buffering effect of guinea pigs on children with ASD in primary school classrooms, which was not evident in the typically developing children they assessed. Meints et al. (2022) found that schoolchildren with and without special educational needs both benefited from a four-week dog intervention by showing lower cortisol levels than the control groups, with stronger effects in the special educational needs' cohort. These results emphasise that different populations do not necessarily gain an equal advantage from the application of the same intervention.

Concerning emotional stability and behaviour in children and young people with special educational needs, Anderson and Olson (2006) found that having a dog in the classroom improved children's emotional stability as well as their attitude towards school and fostered responsibility, respect and empathy in children with crisis behaviour problems. Improving emotional regulation and school-based outcomes not only improved social ability, but had the potential to enhance children's learning in lessons and the wider context. Following AAI with a dog in the classroom, Beetz (2013) found more positive attitudes towards school and positive emotions towards learning. While not aimed at specific educational outcomes, classroom AAI studies have also reported beneficial effects of a dog on classroom behaviour as a whole. Kotrschal and Ortbauer (2003) reported improved behaviour and increased attention focused on the teacher, while Hergovich et al. (2002) reported less behavioural extremes such as hyperactivity. Such behavioural changes can positively impact on other areas of development.

Dogs also seem to facilitate social interactions at all ages and act as a social 'catalyst' (McNicholas & Collis, 2000). It is plausible that these differences may be due to the presence of the dog (as opposed to the novelty of an intervention) as children tend to look longer at and have the longest interaction with the dog, and show more responses to the handler when the dog is in the room (Limond et al., 1997; Prothmann et al., 2009).

AAT, particularly, dog-assisted therapy (DAT) has shown similar benefits, including improved behaviour and increased engagement for children with autism (Silva et al., 2011). Schuck et al. (2015) found that while children with attention deficit hyperactivity disorder (ADHD) showed a reduction of symptoms following cognitive behaviour therapy (CBT), children taking part in canine-assisted CBT showed a greater reduction in symptoms compared to those in CBT without dogs. This may be due to children being more interested in, and therefore engaged in the session with the therapy dog. Martin and Farnum (2002) also noted that, while children's interaction with the therapist decreased in favour of interaction with the dog, better compliance with instructions was seen when the animal was present.

Equine-assisted activities (EEA) and equine-assisted therapy (EAT) often involve versions of therapeutic horseback riding (THR), horsemanship or stable management, and recipients are mostly children of varying ages and diagnoses, but also adolescents and adults. Improvements in empathy and behaviour/self-regulation skills have been shown in children with ASD after EAA (Anderson & Meints, 2016; Gabriels et al., 2015); in motor skills (Gabriels et al., 2015; Voznesenskiy et al., 2016); in attention (Balluerka et al. 2015; Ward et al., 2013); in tolerance (Ward et al., 2013), and in social interactions, social responsiveness or functioning (Balluerka et al., 2015; Bass et al., 2009; Harris & Williams, 2017; Malcolm et al., 2017; Ward et al., 2013). Furthermore, a decrease in severity of autism symptoms (Harris & Williams, 2017; Kern et al., 2011), aggression and hyperactivity, and irritability reduction as well as an increase in social cognition, social communication and word production has also been established (Balluerka et al., 2015; Gabriels et al., 2015; Garcia-Gomez et al., 2013). EAAs have also shown beneficial effects including an improvement in brain functional activity and clinical symptoms of ADHD as measured by the K-ARS questionnaire (Hyun et al., 2016), and quality of life (Kern et al., 2011; Lanning et al., 2014).

Borgi et al. (2016) used EAT for one month and found improvements in social functioning, executive functioning and a somewhat weaker effect in motoric abilities, and Trotter and colleagues (2008) employed equine-assisted counselling (EAC) with at-risk children and adolescents and saw improvements in self-esteem, social stress, aggression and behaviour after 12 weeks of twice weekly sessions.

However, other results from EAAs demonstrated a lack of improvement in children with ASD in communication (Anderson & Meints, 2016; Jenkins & DiGennaro Reed, 2013), systemising behaviours (Anderson & Meints, 2016), or problem behaviour and compliance (Jenkins & DiGennaro Reed, 2013). Likewise, Ewing et al. (2007) found no significant benefits for children with severe emotional disorders after participating in equine-facilitated learning. Davis et al. (2009) saw no improvements in the tested variables for children with cerebral palsy after THR, but some improvement in family cohesion. Randomised controlled trials found no changes in cortisol, post-traumatic stress disorder (PTSD) symptoms or resilience scores in veterans with PTSD after six weeks of equine-assisted psychotherapy (EAP) (Burton et al., 2018), while Johnson et al. (2018) did find significant reduction in PTSD symptoms using THR as EAA with veterans.

It is useful to bear in mind that interventions are also employed with other animals, including aquatic animals (see above), but so far, the evidence base is too unreliable and warrants further research (Marino & Lilienfeld, 2020).

8.3 Theoretical Foundation of Animal-Assisted Interventions

Having glimpsed some of the evidence, we need to ask, Why do humans like animals? Why do we enjoy their company? How can they have beneficial effects on humans? To answer these questions, the following theories will be presented in brief and critiqued in relation to AAI: biophilia hypothesis; attachment theory; social support/buffer theory and biopsychosocial model.

The Biophilia Hypothesis Wilson (1984) sees humans as attracted to animals and suggests that this could explain the positive outcomes following AAI due to satisfying the need for connection to nature as animals are part of nature (for more detail on biophilia, see Chapter 2, this volume). More specifically, interaction with animals can provide a heightened focus of attention and increased engagement in activities during AAI sessions (Limond et al., 1997; Melson & Fine, 2015; Prothmann et al., 2009), resulting in positive physiological effects for the human (Baker et al., 2010; Beetz et al., 2012; Friedmann et al., 2011; Somervill et al., 2009).

Consistent with the biophilia hypothesis, research has shown that people prefer natural environments to built-up man-made ones (Kaplan & Kaplan, 1989). Indeed, a bias towards animals and living things is seen during early development (DeLoache et al., 2011; Mandler et al., 1991; Simion et al., 2008). While biophilia may offer some explanations, it is also limited and underspecified. It cannot predict the complex interplay of biological, social and other factors, nor can it adequately explain why, if love of animals is innate, attraction to animals varies, pet ownership is not consistent across the world, and how different cultures or religious beliefs influence these attitudes (Lawrence, 1994; Herzog, 2014). Attitudes towards animals are not always favourable, with the forensic literature demonstrating the extent of cruelty towards animals by both adults and children (Dadds et al., 2006; Faver & Strand, 2003).

Attachment Theory Bowlby (1969) and Ainsworth and Bell (1970) describe a bond between two individuals which builds over time, with different constellations producing different attachment types and behaviours. This bond, usually between the primary caregiver and the child, is formed during early childhood, but can also change over time (see Chapter 2, this volume). Although the original theory envisaged a caregiver and baby attachment, Rynearson (1978) suggested that pets can act as attachment figures, too. This is supported by Hawkins and Williams (2017), who reported children being attached to their pets (see Chapter 9, this volume, for details). This theory would predict beneficial effects of AAI which have been found in studies with interactions taking place over a number of sessions, allowing a bond to develop between human and animal. Carlisle (2014) reported that children with autism who had a pet dog were attached to the animal and showed increased age-appropriate social skills compared to autistic children without a pet. Interestingly, children with insecure attachment types have been also shown to benefit from social interaction with a dog (Beetz et al., 2011, 2012) in ways which would not be possible with another human (Beetz & McCardle, 2017). Attachment to humans may be different to attachment to pets, perhaps due to the different feelings people have towards pets and humans (Smolkovic et al., 2012). Children with autism may not show the same attachment to parents as neurotypical children, but attachment to animals may provide some support and act as a buffer.

However, attachment theory alone cannot explain the beneficial effects following interventions with an animal after a single exposure. For example, cognitive tasks carried out in an educational environment in the presence of a therapy dog resulted in better compliance (Gee et al., 2009) and faster performance (Gee et al., 2007), but it is unlikely that children were attached to the dog as they spent very limited time together; instead, the presence of the dog may act as support for the child.

Social Buffer/Support Theory proposes that a person provides a sense of social belonging to another person which enhances an individual's quality of life, health and well-being (McNicholas & Collis, 2006), and is thought to act as a buffer during adverse life events. This support includes the perceived and actual help the person receives from the people around them (Cohen & Wills, 1985) (see Chapter 2, this volume). Although initially this theory related to human-to-human support, with respect to HAI and AAI, researchers suggest that the dog–human relationship is similar, with the animal providing comfort and a positive social outlet (Bonas et al., 2000). Animals provide consistent, dependable and non-judgemental companionship (Serpell, 2004, 2010; Wells, 2009). Children who lack social support from adults gained emotional support from their pets (Melson, 2003), indicating their importance in terms of social support (McConnell et al., 2011). In line with this, in AAI, children with insecure/disorganised attachment were seen to gain social support from a visiting therapy dog, but not from a toy dog or a human (Beetz et al., 2012). Ward-Griffin et al. (2018) reported that university students showed improvement in their perceived social support following a one-off intervention session with a therapy dog.

Whilst some findings from AAI can be explained by social support theory, this theory fails to provide a comprehensive account of the neuropsychological or physiological mechanisms involved in such beneficial effects, similar to biophilia and attachment theory. In addition, as with the previous two theoretical frameworks, it is unable to account for external factors such as attitudes towards animals, and cultural differences. A more comprehensive framework is the biopsychosocial model which includes biological, psychological and social factors.

The Biopsychosocial Model is one of the most comprehensive models which allows for the incorporation of multiple factors involved in HAIs during AAI. First proposed by Engel (1978), this model explains how biological measures, and psychological and social challenges are related to health outcomes and to each other (Lehman et al., 2017) (see Chapter 2, this volume). Friedmann and Gee (2017) highlight that the biopsychosocial model is useful to describe and explain how the three different realms interact and can therefore account for the findings reported during HAIs and in AAI sessions. The model furthermore allows for subjective experience to be included alongside objective physiological measures to produce person-centred outcomes.

An increasing number of studies demonstrate the beneficial effect of animals on humans' biological functioning. Spending time with pets has a calming and relaxing effect in terms of physiological responses (Fine & Beck, 2015). Cortisol can be seen as an indicator of stress and is affected by close contact with animals. For example, people had nearly double the level of oxytocin and a reduced level of cortisol when stroking their pet dog (Kertes et al., 2017; Odendaal & Meintjes, 2003). The cortisol-awakening response in children with ASD was found to be reduced by 48% when a dog was introduced into the family and increased again to the same pre-dog level when the *dog* was removed from the home (Viau et al., 2010). In AAI studies, in the presence of a dog, lower levels of cortisol were found for children with insecure attachment following a stressful experience (Beetz et al., 2012). Furthermore, schoolchildren with and without special educational needs showed lower cortisol levels after a four-week dog-assisted intervention when compared to two control groups (Meints et al., 2022), and more relaxed and better behaviour has been observed in classrooms in the presence of a dog (Kotrschal & Ortbauer, 2003). The biopsychosocial model is well suited to explain these results as it incorporates physiological factors such as cortisol and oxytocin levels as well as psychological and social influences on the individual. Psychological factors may include, for example, attachment, self-esteem, loneliness, anxiety or depression, and wider social concepts such as friendships, support groups and attitudes towards pets. So far, this model has best integrated these factors involved in HAIs and during AAI.

It could be argued that the biopsychosocial model may be seen as too person-centred, and other models have been suggested as taking into account even wider interactions at community and society level (for further approaches, see Lehman et al. (2017) and Gee and Mueller (2019)). However, overall, the biopsychosocial model is currently best suited to allow for and explain the complex interplay of factors involved in AAI.

8.4 Risks of AAI, Guidelines and Best Practice

The wealth of research evidence presented here showcases the wide variety of activities taking place with animals, across many different types of settings. While attempting to establish the potential benefits and effectiveness of AAIs (or lack thereof), it is important to have guidelines in place to ensure the well-being and safety of both human and animal participants. The welfare of the animals should not be compromised (Brelsford et al., 2020).

8.4.1 Safeguarding

When working with human populations, it is important that everyone is kept safe, particularly when working with children and vulnerable adults. Health and hygiene protocols should be used and followed to ensure cleanliness and prevent the risk of any illness or infection, with additional protocols in case of accidental injury. Personal preferences, allergies and phobias should also be taken into account. As AAI does not set out to reduce phobias, it should not be used for such purposes unless that is the sole purpose of the session, led by a trained professional (Brelsford et al., 2020). Animals, for example dogs, should be insured appropriately.

Individuals delivering or overseeing AAI should understand the needs of the population they work with, especially where informed consent cannot be obtained directly. Safety checks should be carried out with staff employing AAI. Individuals, particularly those who are vulnerable, should never be left alone with the animal and/or handler only, in cases where the handler is solely responsible for the dog (Gee et al., 2017; Meints et al., 2017). It is recommended that AAI have a handler present who is solely responsible for the animal, in addition to a further member of staff who leads the session-related tasks. As an exception, AATs (and at times AAEs) can be led by the therapist (teacher), who can also be trained to deliver the session with their own animal. In this case, there would not be additional personnel in the room, and the client would be offered a high level of confidentiality during the sessions.

It may be necessary to terminate sessions due to animal welfare needs or adverse human behaviour. Ensuring that there are sufficient staff to manage such a scenario is an important factor in the safe administration of AAI (Brelsford et al., 2020).

8.4.2 Animal Welfare Considerations

Animals involved in HAI should be assessed by a veterinarian to ensure they are in good health and assessed for their suitability to work with specific populations of humans. Animals should have suitable working conditions (i.e. possibility to retreat, food, water and toileting facilities available) with limited working times (see Dog Care Plan (Brelsford et al., 2020) as an example). The environment they will work in should be suitable for their needs as well as those of the human (MacNamara & MacLean, 2017). For dogs, it is recommended that an external assessment is carried out by a dog behaviour specialist to ensure the dog's suitability and welfare are considered before interventions begin (Brelsford et al, 2017). Ideally, other animals should also have an assessment prior to intervention start. Doing so will safeguard the animals' welfare, ensuring they are suited to take part in AAI, which will reduce the potential for distress and accidental injury. It is important that assessments are completed prior to the AAI commencing as an animal's behaviour can change over time or in response to the onset of a medical condition arising. Such assessments should also be carried out on an annual basis and with changes in behaviour and/or health in the animal to ensure continued suitability.

In addition to animals being assessed, dog handlers, who would ideally be solely responsible for their animal, should also be trained in recognising dog body language and distress signalling. This will ensure the animals are not placed in a situation where their welfare is compromised due to feeling distressed or uncertain. If a situation does occur which results in distress signalling (Meints et al., 2017; Meints et al., 2018), the handler should be able to recognise the signals and intervene.

To ensure animal welfare is not compromised, ground rules should be set with participants prior to sessions starting. For dogs, this would include having a safe space where the dog is not disturbed, with no crowding, hugging or kissing the dog (Meints et al., 2017).

Ground rules should be set for any species being incorporated within sessions. Smaller mammals such as rabbits, guinea pigs or hamsters, for instance, are prey animals. This means they may be prone to stress if being handled. It is important for those involved to recognise this, ensuring short sessions and not allowing overcrowding or the distressing of small animals are important factors which should be communicated to those interacting with them.

High standards of welfare and compassion towards animals should be of paramount importance during all activities. Sessions must be stopped if either the animal or the human participant shows any signs of stress or indicates that they are no longer enjoying the interaction (see Brelsford et al., 2020).

8.4.3 The LEAD Risk Assessment Toolkit

How can we ensure that AAI providers and users adhere to the highest human safety and animal welfare standards? It is noteworthy that while many organisations adhere to their own internal guidelines (e.g. Pet Partners, USA; Pets as Therapy, UK, Society for Companion Animal Studies, UK; Animal-Assisted Interventions International, International Association of Human-Animal Interaction Organizations), or to other organisations' guidelines, there is a distinct lack of a set of *unified* guidelines to ensure appropriate risk assessments and welfare standards are adhered to during AAIs (see Serpell et al. (2020) for an overview of the widespread variation of guidelines in the United States).

The development of the Lincoln Education Assistance with Dogs (LEAD) Risk Assessment Toolkit (Brelsford et al., 2020) has closed this gap. The LEAD Toolkit includes a comprehensive and flexible set of guidelines and risk assessment tools which can be used in any setting and can be adapted to incorporate setting-specific protocols and policies (Brelsford et al., 2020). The Toolkit is freely available to download and can be accessed via the link at the end of this section. It was constructed and published as the result of a large-scale randomised controlled trial on dog-assisted interventions with schoolchildren with and without special educational needs (https://lead.blogs.lincoln.ac.uk) and with an international usership of AAI stakeholders in mind (Figure 8.2).

The LEAD Risk Assessment Toolkit contains a risk assessment tool, an animal welfare plan and best practice guidance. The risk assessment can be adapted to any setting employing any type of AAI and can also be adapted to accommodate the inclusion of various animals. In addition to this, and due to the popularity of therapy dogs in school-based sessions, a further tool is available which is specifically designed to assess risk in school settings for dog-assisted interventions.

The toolkit provides an easy-to-use structure to ensure all areas of human and animal safety and welfare in AAI are considered and wider risks assessed. Both versions are adaptable to ensure that relevant policy and practice of each specific setting can be appropriately incorporated. Both versions can be used paper-based or in electronic form.

Figure 8.2 Video by Prof. Nancy Gee discussing the benefit of the LEAD Risk Assessment Tool: (www.youtube.com/watch?v=YpGG X3iSv_Q).

In addition, the risk assessment includes wider considerations such as safeguarding, health, safety and hygiene protocols, which are important aspects of any activities involving human and animal interactions. The LEAD toolkit aims to ensure that animal care and welfare are always considered and not compromised.

Within the LEAD toolkit, the dog care plan is recommended for use alongside the risk assessment and can be adapted for other animals. Animal welfare is ensured through positive interactions and the presence of breaks as required, as well as access to fresh drinking water, appropriate room in terms of space and temperature, and time for the animal away from interactions for rest. As every animal is an individual, these considerations need to be appropriate for each animal. Above all, if an animal is feeling tired or shows signs which indicate that they are not enjoying the interaction, the session should be terminated.

The LEAD welfare and safety considerations are important and should be implemented for all AAI and adapted as appropriate. This is particularly the case as the popularity of animals in different settings increases. This leads to more working animals (usually pets) and, likewise, our responsibility for their well-being increases. Negative interactions will not only have a negative impact on the animal but are also likely to cause a negative impact on the effectiveness of the AAI and may increase the risk of injury. It is paramount that guidelines and safety procedures are understood and followed by all involved.

8.5 Conclusion

From the research presented here, it is evident that studies involving animals cover many different topic areas, are carried out across a variety of environments, and involve different types of populations.

Important factors to ensure much-needed consistency in the field of AAI are the adoption of consistent terminology – to which the IAHAIO's 2014/2018 definition has greatly contributed – and the implementation of universal user and stakeholder risk assessment and adherence to safety and welfare guidelines, including best practice guidance with integrated animal welfare considerations (e.g. the LEAD toolkit). From a methodological viewpoint, the robustness of studies still needs to be improved (Rodriguez et al., 2021). It is crucial here to consider the study or application design. For research studies, randomised controlled trial (RCT) designs are the 'gold standard' to measure intervention effects using control group(s). Unfortunately, in the area of HAI and AAI, there is still a significant lack of RCT studies. Many studies fail to include control conditions; hence, assessment of interventions is often limited. Studies are often not sufficiently powered. That is, the sample size is too small, so it is not possible to conclude with confidence that observed effects are valid or bear clinical significance. This in turn opens up important conversations on whether incorporating animals into such activities is ethical or appropriate. This is important from an animal welfare perspective, in addition to the financial cost for those individuals and organisations.

Many of the benefits attributed to AAI are gained from self-report measures. While standardised self-report questionnaires are very useful, additional physiological measures can enhance our understanding of the mechanisms of how AAI affects humans. Furthermore, at times additional parental, teacher or other feedback within the same study may provide further results as to the effect of the animal on the child.

Little is known about the ideal dosage and structure of AAI sessions, for example the amount of time, or the quality of the interactions which provide optimum benefits. These are important considerations to take into account when planning AAI or when assessing the effects of AAI, and this area warrants further research.

Finally, while longitudinal studies and RCT designs still need to grow in number in this young field to improve the evidence base of HAI and AAI, looking at the progress over the last decade (McCune et al., 2020) allows us to look forward to the development of this fruitful field with optimism.

Additional Resources

The LEAD Risk Assessment Toolkit is freely available to download and can be accessed via:

- Best Practice Standards in Animal-Assisted Interventions: How the LEAD Risk Assessment Tool Can Help (2017). Pdf copy available at: www.mdpi.com/2076-2615/10/6/974
- The LEAD Risk Assessment Tool can be accessed and downloaded for use at: www.mdpi.com/2076-2615/10/6/974#supplementary

Further information can be found here:

- https://lead.blogs.lincoln.ac.uk/
- www.waltham.com/news-events/safe-animal-assisted-interventions-schools

8.6 Learning Outcomes

Source: Illustrated by Loïc Maréchal adapted from istockphotos.

Chapter Summary

- Animal-assisted intervention is the umbrella term, and more specific terms are used when talking specifically about animal-assisted activities, animal-assisted therapy, animal-assisted education, and animal-assisted counselling or coaching.
- The interventions are all goal oriented and planned, but vary in terms of the type of animal included, their purpose, structure and length.
- Before implementing human–animal interaction sessions, detailed guidelines and a thorough risk assessment such as the LEAD tool should be used.
- Further research into the benefits of human–animal interaction and animal-assisted interventions needs to enhance scientific rigour via using randomised controlled trial and longitudinal research, sufficient sample sizes, and detailed testing protocols to ensure treatment fidelity and replication.
- Future research needs to investigate conditions for intervention effectiveness such as optimal session structure and dosage, longitudinal benefits, and who benefits most. This will enable better planning for future interventions.

Check Your Understanding

Multiple Choice Questions

Question 1: Which type of interaction can be provided by a volunteer dog handler with a professional therapist?

- Answer a: Animal-Assisted Activities.
- Answer b: Animal-Assisted Therapy.
- Answer c: Animal-Assisted Education.

Question 2: The LEAD Risk Assessment Toolkit is:

- Answer a: A risk assessment tool.
- Answer b: An animal welfare plan.
- Answer c: Best practice guidance with a risk assessment tool and dog care plan.

Question 3: Which theory suggests that animal-assisted intervention is beneficial due to the person developing a bond with the animal?

- Answer a: social support theory.
- Answer b: attachment theory.
- Answer c: biopsychosocial model.

Open Questions

1 Use the information in this chapter to evaluate and critically analyse the following statement: human–animal interactions are beneficial for human participants.
2 How would you ensure that animal-assisted intervention providers and users adhere to highest human safety and animal welfare standards?

9 Human–Pet Interaction

Ana Maria Barcelos, Daniela Pörtl and Roxanne Hawkins

9.1 Introduction

Companion animals, hereby referred to as 'pets', are animals kept indoors (e.g. dog) or outdoors (e.g. horse) for either pleasure or companionship. For thousands of years, humans have been sharing their everyday lives with pets, the most common being dogs and cats. More than companions, they have reached the status of family members for many and can be an intrinsic factor for one's quality of life, helping to cope with mental health issues or physical mobility difficulties. To understand how pets became and remain so important in human society, a few questions need to be answered. How did wild animals turn into companions? How can one describe the human–pet relationship and how pets are perceived? What are the physiological, psychological and physical similarities between humans and pets? The three sections of this chapter help answer these inquiries. Focusing on biological, psychological and social explanations, this chapter seeks to understand similarities between humans and companion animals through the biopsychosocial model, weighing the contribution of each individual factor to explain human health (Engel, 1977). The interaction between these factors has recently been proposed to explain how human–pet relationships affect human health or well-being (Gee et al., 2021). Importantly, this chapter emphasises dog–human interactions (rather than other pet–human interactions), both because dogs are the most common household pets worldwide (Pet Food Manufacturer's Association (PFMA), 2021), and because research on other species is not always available.

9.2 Pet Domestication

Domestication of animals and plants is one of the most remarkable cultural achievements in the evolution of modern humans. We eat domesticated plants; we are surrounded by farm animals selected for wool, milk, egg and meat production; and we enjoy pet companionship. The *Cambridge Dictionary* defines domestication as 'the process of bringing animals or plants under human control in order to provide food, power, or company'.

Domestication of animals is the selection for tameness and docility not only of an individual but of a whole species. However, the selection for tameness is accompanied by unexpected traits not observed in their wild forebears including: changes in coat colour such as white patches; smaller teeth and jaws; relatively smaller brains; changes in craniofacial morphology, floppy ears, curly tails; altered female sexual cycles and prolonged juvenile behaviour; alterations in adrenocorticotropic hormones; and changed concentrations of several neurotransmitters (Trut et al., 2009; Wilkins et al., 2014).

DOI: 10.4324/9781003221753-9

Every domesticated species differs in the number of these exhibited traits. Therefore, the term 'domestication syndrome' has been used increasingly in recent studies (Hare et al., 2012; Pörtl & Jung, 2017, 2019; Wilkins et al., 2014, 2021). The term 'syndrome' describes the variability of traits which should be exhibited to some degree but do not have to be exhibited completely in every domesticated species. The term domestication syndrome is not a hypothesis but a generalisation from observations (Wilkins et al., 2021; Wright et al., 2020). Using dogs as an example, the domestication process will be explained.

Dogs appear to have been the first domesticated non-human animal, beginning about 30,000 years ago (Germonpré et al., 2009; Ovodov et al., 2011), followed by farm animals. Dogs are also one of the few species exhibiting all traits of domestication syndrome. Therefore, we take a closer look at dog domestication hypotheses to better understand domestication processes in general.

The first hypothesis of dog domestication supported by Lorenz (1967) and Zimen (1997) considered human hand-reared wolf pups as dogs' ancestors. However, even hand-reared wolf adults pose risks to humans (Kubinyi et al., 2007) and likely left human companionship to seek mates. Therefore, domestication by rearing and taming single animals seems implausible. Coppinger and Coppinger (2001) argued that waste around early human settlements built a new ecological niche where the first dogs originated from. However, it is commonly agreed that dog domestication had already begun in the Upper Palaeolithic, long before the era of agriculture and farming started in the Neolithic. In addition, archaeologists claim that waste dumps are a characteristic of modern times (Havlíček, 2015). Further, archaeologists have found many dog–human graves from the Palaeolithic all over the world (Morey & Wiant, 1992). Dog graves and suspected intensive human care suggest early dogs were not just scavengers but also companions (Janssen et al., 2018). Therefore, animal domestication seems to be a story of genetic isolation due to companionship with humans. Hare (2012) suggests 'surviving of the friendliest' as a model of self-domestication in dogs, with prosocial behaviour as the driving force. Another recent model of dog domestication and domestication processes in general is the 'active social domestication' hypothesis (Pörtl & Jung, 2019, 2017). This model focuses on the impact of interspecific prosocial contacts reducing environmental stress conditions as a kind of domestic niche. This hypothesis considers that (dog) domestication is not exclusively due to artificial genetic selection but additionally an epigenetic-based process focusing on 'environmental programming'. Epigenetic means altered gene expression due to changed environmental conditions (Waddington, 1953), especially stressors and diet.

Each domesticated species is expected to have experienced two semi-distinct domestication stages likely involving different epigenetic and genetic changes (Pendleton et al., 2018; Zeder, 2015). As a form of a natural genetic selection process, the initial stage is suspected to have mainly involved physiological changes that became genetically fixed due to epigenetic modulations (Wilkins et al., 2021). For example, ancient wolves with genetically predisposed friendly behaviour (Kis et al., 2014; Oliva et al., 2016) are suspected to have been able to communicate with humans because both species are highly social and cooperative mammals skilled with similar social communication gestures like complex mimicry and joint attention (Chapter 3, this volume). Living in family groups, both can form close individualised emotional bonds (Berns et al., 2012; Marshall-Pescini et al., 2017; Range & Virányi, 2011). Interspecific communication reduced stress in both. Stress activity is known to be crucial for shaping behaviour and brain function, often with long-lasting effects (Hunter et al., 2015). Limbic brain regions are important for mood control (LeDoux, 2012; Lim & Young, 2006), and they are the central organs of stress and stress adaptation because they perceive threats and initiate behavioural and physiological responses. Mammal domestication had an impact on limbic brain structures. The amygdala, involved in fear processing, reduced in volume, and the hippocampus and medial frontal cortex, both involved in prefrontal fear inhibition, increased in volume (Brusini et al., 2018). Therefore, domesticated animals exhibit reduced fear, decreased reactive aggression and increased inhibitory control, and improved social learning capability.

Changes in the interactions of the hypothalamic-pituitary-adrenal (HPA) stress axis, responsible for the fight and flight response, and the cross-regulated serotonin and oxytocin calming system in the limbic brain (LeDoux, 2012), are expected to be of particular relevance regarding domestication processes (Figure 9.1). Remodelling a domestication process (Trut et al., 2009, 2021), the results of the Siberian farm fox experiment, where silver foxes had been selected for tameness towards humans since 1952, confirmed changes in the adrenal cortex, serotonergic and limbic systems related to a downregulation of the HPA axis within only a few generations compared to the control group foxes (Trut et al., 2009), as well as changes in cerebral epigenetic patterns (Herbeck et al., 2017; Jensen, 2015). Mental skills, like using basic human communicative gestures and better social problem-solving skills, increased in experimental fox kittens (Hare et al., 2005) during the domestication process. These results correspond to epigenetic modulation in the limbic brain due to increased social affection (Buschdorf & Meaney, 2015; Meaney & Szyf, 2005). The active social domestication hypothesis considers that epigenetic

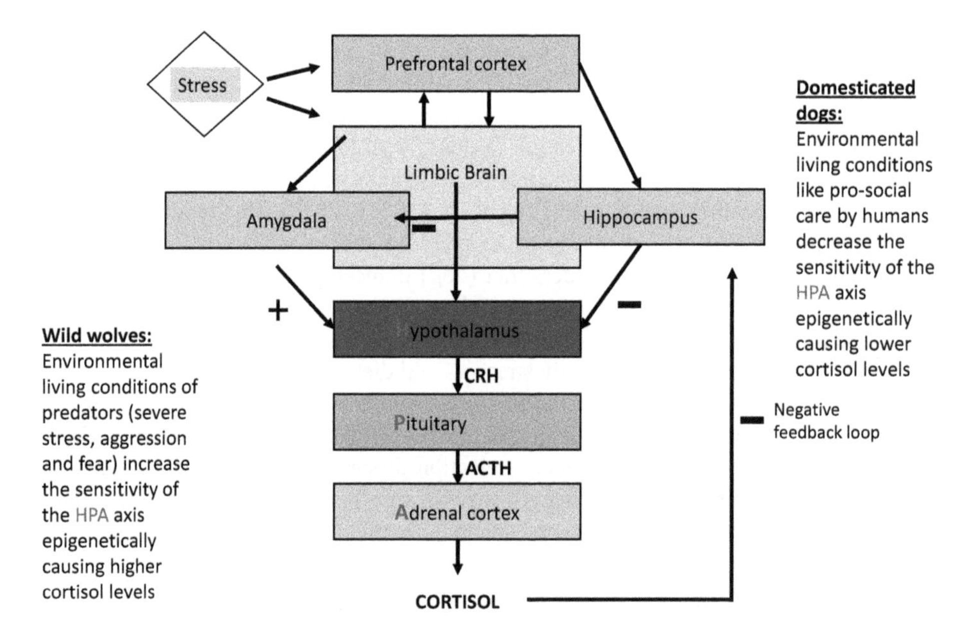

Figure 9.1 Anatomical connections between limbic brain areas (hippocampus/amygdala), the prefrontal cortex and the hypothalamus facilitate activation of the HPA axis. The HPA axis is influenced through an enhancement of the amygdala and an inhibition of the hippocampus. The sensitivity of the HPA axis is programmed epigenetically due to (early) life stress. Under physiological conditions, the HPA axis and the serotonin and oxytocin calming systems in the brain are cross-regulated. We consider that domestication is essentially due to epigenetic programming changing the sensitivity of the HPA axis mediating stress behaviour and its interaction with the brain's calming systems.

downregulation of the HPA axis due to increased interspecific mutual care might be an important mechanism contributing to the domestication syndrome (Herbeck et al., 2017; Jensen, 2015). Furthermore, the 'neural crest domestication hypothesis' (Wilkins et al., 2014) suggests the set of unexpected traits accompanying the selection of tameness, the so-called domestication syndrome, to be the result of a mild neural crest cell migration deficit during embryonic development where migration defects are particularly important. Neural crest cells are a cell group that migrate throughout the embryo and differentiate into a diverse range of cell types. Given that many of the tissues involved in the traits of domestication syndrome derive from the neural crest, these traits can be interpreted by a reduction in the number of embryonic neural crest cells that give rise to these structures like skin, pigmentation, sympathetic response and the reproductive cycle. Altogether, the domestication process of dogs altered the developmental pathway, enhancing dogs' cooperative communication with humans (Salomons et al., 2021).

The second, much longer and still ongoing stage of domestication is the breed formation period caused by humans selecting animals deliberately in closed genetic populations for special traits. Artificial selection has sculpted the size, shape and behaviour of the modern domesticated dog for at least 14,000 years (Akey et al., 2010). Large numbers of genetic changes took place in this period, overlapping with the initial genetic alterations (Ostrander et al., 2019; Trut et al., 2021; Wilkins et al., 2014). Relevant differences in dog breed behaviour are highly heritable (MacLean et al., 2019). Furthermore, compared to wolves, dogs exhibit changes in genes that are important for brain function as well as in genes with a key role in starch digestion (Axelsson et al., 2013; Saetre et al., 2004). The diet adaptation in dogs reflects the spread of prehistoric agriculture (Arendt et al., 2014, 2016), as a form of convergent-evolution with humans. Further, hypersociability, as a core symptom of dog domestication, is associated with structural gene changes in dogs that are linked to the Williams–Beuren Syndrome in humans, likewise accompanied by hypersociability (von Holdt et al., 2017). To sum up, we can define pet domestication as a syndrome and a still ongoing process of several stages based on mutual interspecific care and companionship with humans instead of a completed state carried out by humans alone. All traditional domesticated animals are almost certainly expected to have experienced these two semi-distinct stages during domestication, as we have stated for dogs. The initial stage of habituation to human presence increased tameness and docility and reduced reactive aggression, involving physiological and epigenetic changes that became genetically fixed followed by the much longer 'breed formation' stage (Wilkins et al., 2021).

9.3 Human–Pet Attachment and Beliefs in Animal Minds

There are several complex interrelated human psychological factors that can affect the nature of human–pet interactions. Attachment or a 'bond' with a pet animal is one such important affective dimension (Figure 9.2). Evidence suggests that people of all ages form strong, emotional connections to their pet animals, viewing them as an integral, and often focal, part of their families and social networks. Children and adolescents often view their pet as their best friend, and report similarities between their relationship with their pets and their siblings, including features such as satisfaction, disclosure and companionship, but less conflict (Cassels et al., 2017). Pets offer people an unconditional and stable sense of comfort, acceptance, loyalty and emotional support and reassurance, being non-judgemental, satisfying the basic need to be loved and to feel safe, and providing intrinsic rewards rather than extrinsic support (Sable, 2013).

Although attachment research usually focuses on childhood, separation distress, including that toward pets, is important throughout the entire lifespan. Attachment theory (see Chapter 2, this volume) can be used as a conceptual lens and framework for examining individual differences

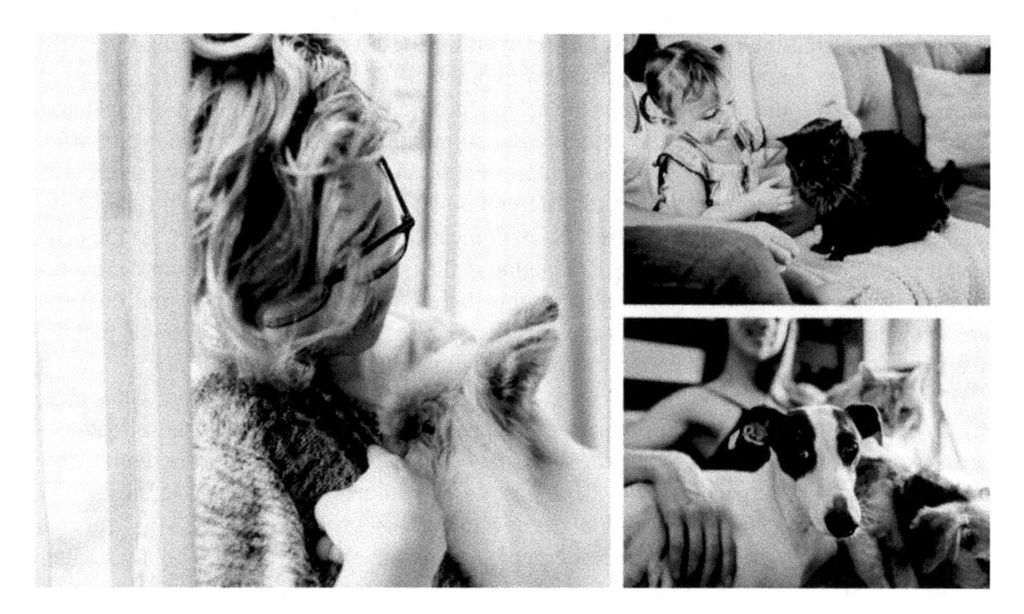

Figure 9.2 Images displaying the human–pet bond.

Source: Helena Lopes (@wildlittlethingsphoto) and Chewy @chewy from https://unsplash.com/.

in human–pet relationships. Research suggests that human–pet bonds are consistent with human attachment theory and that they share the same attachment features (proximity seeking, safe haven, secure base, separation distress) as seen in human–human attachments (Zilcha-Mano et al., 2011, 2012). Proximity seeking, that is, a desire to be close, is one of the most notable features of human–pet attachment (as found with dog owners: Kurdek, 2009). However, pets may not fulfil all functions of an attachment relationship. For example, separation distress observed in pet attachments, may reflect concerns for the pet's welfare more than feeling less safe in their absence (Kwong & Bartholomew, 2011). Moreover, similar negative features of attachment can also be seen in human–pet attachments including insecurities, anxieties, avoidance and negative expectations.

As with human relationships, there are individual differences and different dispositions in experiences of human–pet attachments, each varying in warmth, closeness, emotional involvement and commitment (Zilcha-Mano et al., 2011, 2012). Individual differences in attachment orientations are usually measured along two orthogonal dimensions. These are attachment-related anxiety associated with hyperactivity of the attachment system, and attachment-related avoidance associated with the deactivation of the attachment system (Mikulincer et al., 2003). Those who score low on both dimensions are said to have a secure attachment. Such individuals tend to have positive internal representations of relationships, feel secure and have trust in their relationships, and have a positive image of others with an expectation that others will be accepting and responsive. Those who score high on both dimensions are said to have an insecure attachment, holding negative internal representations of relationships, viewing others as unavailable frustrating or untrustworthy.

As briefly mentioned in Chapter 2, these dimensions have been applied to human–pet attachments. Some studies have identified similarities in insecure human–human and human–pet attachments (Dowsett et al., 2020), though others report less avoidance and anxiety in attachment to dogs (Green et al., 2018). Zilcha-Mano et al. (2011) found that avoidance in human–pet attachments manifested through a need for emotional and physical distance from pets, and discomfort with closeness and intimacy, along with difficulties in depending on a pet for emotional support and comfort when distressed. Anxiety in human–pet attachments is manifested through intense and intrusive worries about pets, a strong desire for proximity, constant reassurance seeking from the pet, and intense frustration (for example, when the pet is displaying affection for another person, or when the pet relationship is not as close as the person desires it to be). Differences in strength and type of attachment may also depend on the pet type. For example, Hawkins et al. (2017) found that children in the United Kingdom with pet dogs scored the highest on pet attachment, followed by children with cats, small mammals (e.g. hamsters, guinea pigs), other (e.g. horses), fish and reptiles, amphibians and, finally, birds. Most pet attachment research, however, focuses on attachment to dogs, possibly because dog ownership is common across cultures, whereas keeping pets of other species (e.g. rabbits) may only be present in specific countries or cultures.

It is well established from human attachment research that secure attachments are important for healthy psychological development, protecting against psychopathology. Similarly, emerging evidence has found that secure human–pet attachments have been associated with a range of positive health and psychosocial outcomes across the lifespan (Gee & Mueller, 2019; Purewal et al., 2017). Benefits of pets for health and well-being have been found for a variety of species including smaller mammals (Muldoon et al., 2019), although most research points to benefits derived from dogs (Endo et al., 2020). Pets may be an important bond in the absence of secure human attachments or attachment anxiety with humans, acting as a buffer where there is a lack of positive human social connection and social support (Chapter 2, this volume; Smolkovic et al., 2012). Human–pet attachments can therefore serve as a substitute for other interpersonal relationships and may be particularly pertinent in cases of adversity (such as childhood exposure to animal and human abuse; Hawkins et al., 2019a). Pets may also offer a pathway toward re-establishing attachment security with other humans (such as for children within foster care, Carr & Rockett, 2017).

There seem to be mutual benefits of a secure attachment to a pet, such as health and well-being outcomes for the pet owner, and increased welfare and care for the pet. However, as with human interpersonal relationships, insecure pet attachment may negatively impact upon the nature of human–pet interactions and pet-directed behaviours (Green et al., 2018). For example, Hirschenhauser et al. (2017) found that avoidant child–pet attachment styles were associated with less time communicating with pets and avoidance of a caring relationship. A lack of human–pet attachment, or insecure attachment style, may also pose a risk for acceptance of animal cruelty and animal cruelty behaviour (Hawkins et al., 2017, 2020). Given the importance of attachment types for human–pet interactions, this field needs to continue to grow. An important research avenue will be examining whether positive pet attachment can be facilitated, such as through intervention or education.

Arguably one of the most important cognitive mechanisms underpinning humans' interactions with, and attachment or bond to, their pet, is whether owners believe that their pets are sentient beings, that is, can experience their own thoughts and emotions and have consciousness. In the human–animal literature, this psychological factor has been labelled a 'belief in animal mind' (BAM: Hawkins & Williams, 2016) and can be broadly categorised as beliefs in (1) animals having cognitive abilities (e.g. can problem solve) and (2) animals having emotional capabilities (e.g. can experience joy and fear). In Chapter 2 (this volume), the expression of emotions by animals is discussed. Belief in animal mind relates to empathetic skills and 'emotional attunement', that is, an understanding of and knowledge about emotional capabilities and perspective-taking (cognitive empathy), and an ability to share, attune to or resonate with the feelings of others (affective empathy). Such beliefs closely relate to 'anthropomorphism', which means the attribution of human-like characteristics and mental states to nonhuman animals (Serpell, 2003; see Chapter 2, this volume, for further detail about anthropomorphism). According to social psychology and attribution theory, humans may explain pet behaviour (e.g. chewing furniture) through attributing the causes of behaviour to internal states of the pet (e.g. loneliness, boredom), which can impact upon how that pet is viewed and treated (Sanders et al., 1999). Psychological explanations of pets' behaviour by humans may derive from a person's desire to socially engage with the animal and place value onto them, infer ownership of the pet, and may identify with the pet and have a need for psychological closeness and affiliation (Caporael & Heyes, 1997; Kiesler et al., 2006). These attributions may also relate to self and reflexive consciousness and the development of social skills, for example, the ability to make inferences about the mental experiences and behavioural intent of others based on one's own self-awareness and understanding of self in the environment that become more complex with age (i.e. theory of mind: Baron-Cohen et al., 2013).

Research has shown that most 'lay' people believe in the emotionality of pet animals, but these beliefs, and confidence in these beliefs, are more prominent in pet owners (Knight et al., 2004) and are especially prominent in dog owners, compared to other types of pet owners (e.g. fish). For example, dog owners commonly believe that their animals feel guilt after doing something wrong. However, Horowitz (2009) disentangled this in an experiment that actually indicates that owners' own behaviour upon the dog's potential wrongdoing is what triggers the 'guilty look' response in the dog, rather than this being a real manifestation of guilt. Pet owners' beliefs in their animal's emotionality can possibly be explained through 'the contact hypothesis' and a 'sense of oneness' with the pet. For example, an understanding of pets' emotions may be developed over time through spending quality time with the animal, sharing positive experiences with the animal, and making direct observations of the animal's emotional and behavioural expressions (Maust-Mohl et al., 2012), which is only possible with some types of pets. Therefore, it is not pet ownership alone that is important for the understanding of, and belief in, animal minds. A close relationship with, affection towards, and felt emotional attachment towards the animal, can result in more psychological attributions of emotions and explanations for observed animal behaviour (Kiesler et al., 2006). The attribution of primary emotions (e.g. happiness, sadness, fear, surprise) is more commonly observed than the attribution of complex, secondary emotions (e.g. love, jealousy, shame, grief) (Martens et al., 2016), possibly because secondary emotions have been viewed as restricted to humans and other primates (Morris et al., 2008). In relation to emotional and cognitive capabilities, humans tend to rate animals based on familiarity with, phylogenetical closeness to, and/or degree of attachment bond to a particular animal (Borgi & Cirulli, 2015; Eddy et al., 1993). Dogs tend to receive the highest rating for sentience compared to other pet types (e.g. birds) by both adults and children across countries, possibly due to reciprocal human–dog attachments, complex and varied human–dog interaction types, and the perception that dogs possess similar mental processes to humans (Menor-Campos et al., 2018).

Believing in, and understanding of, animal sentience is arguably at the core of human–animal interactions. Such beliefs can impact upon the ability to process non-human emotions as well as impacting upon feelings and attitudes towards pets, and therefore subsequently the strength of the human–pet bond (Urquiza-Haas & Kotrschal, 2015). Beliefs about animal minds can also impact upon people's views towards animal rights and the uses of animals, pet-directed behaviours and concerns regarding pet welfare across cultures (Cornish et al., 2018). A lack of belief in, or understanding of, pet sentience may increase risk for negative attitudes towards the treatment of pets and, potentially, animal cruelty behaviours (Hawkins & Williams, 2016). Therefore, increasing such beliefs is an important endeavour to improve the welfare and treatment of pets. Beliefs about animal minds can be influenced through improving knowledge about the emotional and cognitive capabilities of animals through both informal learning (e.g. experience with pets) or more targeted instruction (e.g. education programmes), which may have positive influences on human–pet interactions. Increasing understanding of animal sentience and recognition of emotional signals in a variety of pet types has therefore been at the core of animal welfare interventions and will continue to be an important avenue for future research (Hawkins & Williams, 2019).

9.4 Pet–Human Similarities in Oxytocin, Cortisol, Personality and Physical Attributes

One of the questions that frequently intrigues those who live or interact with pets is whether pets are similar to them, not just physically but also at the psychological and the physiological level. This section covers four of the main factors investigated in the literature of pet–human similarities: oxytocin, cortisol, personality and physical characteristics.

9.4.1 Oxytocin in Pets and Humans

Oxytocin, also called the 'love hormone', is a hormone and neuropeptide synthetised in the hypothalamus and released into the bloodstream by the pituitary gland of mammals. Four of several well-documented roles of oxytocin are: 1) reproductive – womb contraction during labour, milk ejection in lactation; 2) social – bonding facilitation (e.g. between mother and infant; dog and owner), affiliative, maternal and prosocial behaviour, social learning/feedback facilitation; 3) psychological – reduction of stress and fear; and 4) cardiovascular – vasodilation and decrease in blood pressure (Powell et al., 2019; Thielke & Udell, 2017). The social role of oxytocin has received plenty of attention in the literature of human–pet interaction, as scientists seek to better understand the underlying function of oxytocin in the human–pet bond, human–pet communication and in health benefits obtained from human–pet interactions. However, it is important to note that the concentration of oxytocin usually measured in pet–human studies, called peripheral oxytocin (e.g. from the urine, saliva, blood), does not always correlate with the concentration of central oxytocin (e.g. collected from the cerebrospinal fluid, Valstad et al., 2017). For example, in humans under stress, there is a correlation between these concentrations, whereas under basal conditions, they seem not to be correlated (Valstad et al., 2017). Further studies are needed to disentangle the central-peripheral relationship in pet–human interactions.

Odendaal (2000) documented, for the first time, changes in human oxytocin levels due to human–animal interactions. Human participants, after positively interacting with their own dogs, displayed a greater increase in oxytocin than people in two control groups: positive interactions with an unfamiliar dog and reading a book quietly. Since then, human–dog interactions (e.g. touching, looking at, speaking to) have been repeatedly reported, in the literature, to increase oxytocin in both humans and dogs. Interestingly, there are several factors that seem to be associated with oxytocin release in human–dog interactions, such as gender (higher oxytocin concentration in female humans, Miller et al., 2009), type of human–dog interaction (e.g. long dog-to-human gaze induces greater oxytocin release in humans than short gaze, Nagasawa et al., 2009), dog characteristics (e.g. the breed, Tonoike et al., 2016), and familiarity with the dog (i.e. contributes to greater oxytocin concentration in humans, Odendaal, 2000), among others. Human oxytocin release also differs depending on the species of the animal they interact with. Although little evidence is available on other companion animal species (non-dog), studies on human–cat interaction suggest that the general exposure to a cat does not increase oxytocin concentration in humans (Curry et al., 2015; Johnson et al., 2021), but like in dogs, the type of human–cat interaction also influences hormonal release. Petting the cat gently or the cat initiating contact with the human, for example, is positively correlated with oxytocin levels, while the human initiating the contact or playing with the cat are not correlated (Johnson et al., 2021).

An oxytocin-mediated positive loop between humans and animals occurs when an increase in oxytocin concentration in one of the two species leads to increased oxytocin in the other and so on. For example, Nagasawa et al. (2015) found that higher levels of oxytocin in dogs increased gazing behaviour in dogs, which enhanced urinary oxytocin concentration in owners. The same positive loop did not occur between humans and wolves. Based on their results, it has been suggested that human–dog convergent evolution is involved in oxytocin increase in such circumstances. However, recent evidence has shown that dogs and wolves living in enclosures, after human interaction, display similar levels of oxytocin (not significantly elevated), while pet dogs have a larger increase in oxytocin concentration. These findings might indicate that the life experience of a dog, rather than domestication, plays a bigger role in dog oxytocin release (Wirobski et al., 2021).

9.4.2 Cortisol in Pets and Humans

The synthesis of cortisol, a steroid hormone, is controlled by the hypothalamic-pituitary-adrenal (HPA) axis (Figure 9.1). In response to stress, the hypothalamus secretes a corticotropin-releasing hormone (CRH), which triggers the pituitary gland to secrete adrenocorticotropic hormone (ACTH), and ACTH stimulates the production of cortisol by the adrenal cortex (Smith & Vale, 2006). In studies of human–animal interaction, cortisol concentration (e.g. in saliva, urine, blood, hair) is frequently measured to assess physiological stress in humans and/or animals. A phenomenon called emotional contagion occurs when the stress level (or another emotional state) of one individual automatically changes in a similar way to what is observed in another individual. In another words, emotional contagion is the tendency to experience another individual's emotions (Hatfield et al., 1992), such as stress.

To date, at least two studies have evidenced cortisol synchronisation in human–pet dyads. Buttner et al. (2015) measured salivary cortisol levels of dogs and their handlers before and after an agility competition, and found that the cortisol of the dyad increased in a synchronised manner, indicating contagion of arousal. Inferences of stress contagion were not directly made because cortisol levels also elevate in exciting situations or under physical exertion (Buttner et al., 2015). More recently, Sundman et al. (2019) found, for the first time, long-term stress synchronisation between humans and animals. Hair cortisol concentrations of dogs and their owners were highly correlated during summer and winter months, and the personality traits of the owners affected dog cortisol levels significantly. Sundman et al.'s (2019) findings indicate that dogs mirror the stress of their owners. Cortisol synchronisation due to physical activity, rather than stress, was ruled out, as dogs' physical activity was not correlated to dogs' cortisol levels.

Lack of short- and long-term cortisol synchronisation between owners and their dogs has also been reported in the literature (Dreschel & Granger, 2005; Höglin et al., 2021). A number of factors have been suggested to interfere with dog/human cortisol levels, such as owner personality (e.g. dogs of more neurotic owners have lower hair cortisol concentration, Sundman et al., 2019), human gender (in the context of agility competitions, male handlers' dogs show greater increase in cortisol than females' dogs, Buttner et al., 2015), dog breed (ancient or solitary hunting breeds do not show long-term cortisol synchronisation with their owners, as oppose to herding breeds, Höglin et al., 2021), dog sex (female dogs display higher cortisol concentration and stronger synchronisation with humans, Sundman et al., 2019), and dog lifestyle (competing dogs' cortisol synchronise more strongly with humans than pet dogs', Sundman et al., 2019). Season and dog-owner relationship also seem to be important factors. Dog hair cortisol concentration is higher during winter (Roth et al., 2015), and higher relationship cost perceived by the owner is associated with higher dog cortisol levels (Höglin et al., 2021). Future studies should consider these factors and investigate other pets, as information about non-dog pets is scarce.

9.4.3 Human and Pet Personality Traits

Personality traits can be defined as relatively stable characteristics of an individual's behaviours, thoughts and feelings across different circumstances (Roberts, 2009). In humans, personality traits are most often categorised according to the Big Five Model (Chapter 5, this volume). In this model, a person is ranked on five traits: (1) extraversion (being sociable, energetic, warm), (2) agreeableness (being helpful, trusting, empathetic), (3) conscientiousness (being self-disciplined, organised, thoughtful), (4) neuroticism (being insecure, anxious, unhappy) and (5) openness (independent, imaginative, with many interests, John & Srivastava, 1999). Models of personality traits in companion animals, such as dogs, rabbits, cats and horses, have also been proposed in the literature. However, the structure of personality classification commonly differs from one species to another. The Dog Personality Questionnaire (DPQ: Jones, 2008), for example, assesses the following five dimensions: aggression towards people, aggression towards animals, fearfulness, activity/excitability and responsiveness to training. Comparisons of personality are more commonly observed between individuals of the same species, but there is a small body of research on cross-species personality, mainly investigating similarities and differences between pets and humans.

Turcsán et al. (2012) claim to have conducted the first study to compare the personality of dogs and their owners. They found reliable and significant positive correlations between dogs and owners in four of the five personality dimensions (neuroticism, extraversion, conscientiousness and agreeableness), suggesting that dogs are like their owners. A more recent study also evidenced great personality compatibility between dogs and their owners (Chopik & Weaver, 2019). For example, fearfulness in dogs was associated with neuroticism in owners, more active/excitable dogs had more extroverted owners, and higher responsiveness to training in dogs was linked to conscientious owners. Interestingly, in a study with 3,331 cat owners, as mentioned in Chapter 5 (this volume), higher neuroticism in owners was also linked to problematic behaviours in their cats, such as more aggressive and fearful/anxious behavioural styles (Finka et al., 2019). Chopik and Weaver (2019) suggest at least three main factors that could contribute to similarity in pets' and owners' personalities: 1) the selection of a companion animal that matches the characteristics of the owner, 2) the shared environment and activities of the dyad, 3) the owner's perception of likeable things (i.e. the dog) as similar to themselves. The last, however, is less likely to explain the findings of dog-owner personality matching, because the literature indicates that non-owners and owners rate the personality of a dog similarly. Finally, pet–owner personality match is not just an interesting finding; it can have large implications for pet adoption success and relinquishment, as it can predict owner satisfaction with their cat (Evans et al., 2019) and dog relationship (Curb et al., 2015).

9.4.4 Physical Similarities in Pets and Humans

Do pets resemble their owners? One of the classical investigations in this area was conducted by Roy and Christenfeld (2004), who asked judges to match photographs of 45 owners with pictures of their dogs. The authors proposed two mechanisms for the potential pet–owner physical match: 1) people choose an animal that is similar to them, or 2) the features of the person and animal converge over time. In case the first mechanism was correct, they expected that purebred dogs and their owners would show greater similarity than non-purebred and their owners, because the final appearance of a non-purebred puppy is unpredictable. In case the second mechanism was correct, they expected to find a positive correlation between length of dog ownership and matching success. The study found above-chance matching of purebreds and their owners, but the same did not apply for non-purebred dogs; and length of dog ownership was not correlated with matching success. These findings indicate that pets can resemble their owners and that people seek a pet that is similar to them (Roy & Christenfeld, 2004). However, not long after this study, Levine (2005) strongly criticised Roy and Christenfeld's (2004) data analysis and results. This resulted in a new investigation with 96 new naive judges, who assessed 24 dog/owner pictures (Roy & Christenfeld, 2005). Their new results showed again that the matching of purebred dogs and their owners was above chance, supporting their original claim – purebred dogs resemble their owners. Other investigations conducted in different countries (e.g. Venezuela, Japan) provided further evidence for the dog–owner physical resemblance theory (Nakajima et al., 2015). Therefore, it is likely that dog owners really choose dogs who look like them (Figure 9.3).

More specific factors related to the physical similarities between pet owners and their animals have also been investigated. Nakajima (2015) tried to identify the critical feature underlying dog–owner facial similarity. Participants were asked to assess sets of photos of real and fake dog–owner pairs. As expected, they matched the correct pairs more often than the fake ones. This also occurred when they were able to see only the eye region of the owners and the dogs, or when just the mouth region of the owners' faces was masked. In contrast, when the eye region of either

Figure 9.3 Do these dog owners resemble their dogs?

Source: Sophie Davies, Marcus Vinicius Fragoso, Sophia Liu, Kyenner Oliver.

the dogs or the owners was masked, the performance of the participants decreased to around 50%. These results indicate that the eye region is a critical feature of dog–owner resemblance. Another study, in the United States, investigated whether the body mass index (BMI) of 38 dog owners was associated with their dog's body composition score (BCS) (Linder et al., 2021). A strong correlation between dogs' BCS and owners' BMI ($r = 0.60$, $p < 0.001$) was found, indicating dog–owner body similarity. In the Netherlands, Nijland et al. (2010) also found an association between being overweight in dogs and their owners, but not between cats and their owners. Finally, findings of pet and owner resemblance suggest not only that people select animals which are similar to them, but also that people's lifestyles could influence how their animals look, which may have negative (e.g. overweight dogs) or positive (e.g. healthy dogs) consequences.

9.5 Conclusion

The domestication of pets, human–pet attachment, belief in animal mind (animals as sentient beings), and the similarities humans and pets share are all interconnected. The second stage of dog domestication (breed formation period), as an ongoing process, illustrates this well. People actively select animals according to traits they are interested in, which could reflect positive experiences or secure attachment they had with a specific animal/breed in the past, or could reflect people's search for an animal that will match their lifestyle (e.g. an active person looking for a dog that endures long runs). Furthermore, the resulting relationships people have with their selected pets can determine whether they see animals as sentient beings or not and could also impact on human–pet hormonal synchronisation. Finally, pet–human relationship is a complex subject that deserves further investigation, especially regarding its benefit/detriment to animals, as they are passively chosen by humans, and do not actively choose who they live with.

9.6 Learning Outcomes

Source: Illustrated by Loïc Maréchal adapted from istockphotos.

Chapter Summary

- Understanding pet domestication as a syndrome and as an ongoing process based on interspecific prosocial communication and companionship rather than a state.
- An understanding of some of the cognitive and affective psychological mechanisms underpinning human–pet interactions including attachment and beliefs about animal minds.
- Human–pet interaction goes far beyond companionship, and pets and humans are physiologically, psychologically and physically connected, resembling each other in various aspects.

Check Your Understanding

Multiple Choice Questions

Question 1: In most mammals, cortisol can cross the placenta. To what extent might maternal cortisol levels influence embryonic neural crest cell migration deficits, resulting in the 'domestication syndrome'?

- Answer a: Maternal cortisol level is the main factor underlying this cell migration.
- Answer b: It is still an open question.
- Answer c: Maternal cortisol does not influence embryonic neural crest cell migration.

Question 2: According to Zilcha-Mano et al. (2011), which is a feature of an anxious human–pet attachment?

- Answer a: Emotional distance.
- Answer b: Secure feeling.
- Answer c: Intrusive worries.

Question 3: Based on recent findings (Wirobski et al., 2021), what is the main factor responsible for high oxytocin release in dogs interacting with humans?

- Answer a: Dog domestication.
- Answer b: Life experience.
- Answer c: Type of interaction.

Open Questions

1. What makes human relationship with pets special? Think about the process of pet domestication, the way pets are seen by humans, and the characteristics humans and pets share.
2. How might owner personality affect the welfare of their animal? Please give two examples.

10 Human–Livestock Interaction

Grahame J. Coleman and Paul H. Hemsworth

10.1 Introduction

Domestication of livestock dates back over 10,000 years (Vigne, 2011). However, despite the fact that domesticated animals have been dependent on their human carers and that humans are a critical part of domesticated animals' social environment, the scientific study of human–animal interactions (HAIs) has only recently seen a marked increase. Nevertheless human–livestock interaction is still underrepresented within HAI research. Out of 1,715 articles reported in the Web of Science between 1982 and 2018: 'livestock' was mentioned in 100 articles (5.83%), pigs were mentioned in 90 articles (5.25%) and chickens in 34 articles (1.98%), while dogs or other companion animals (41.16%) were by far the most commonly represented animals (Yatcilla, 2021; Chapter 1, this volume).

The media frequently reports cases of mistreatment of animals in abattoirs and in the livestock industries in general (Rice et al., 2020). While the actual prevalence of such mistreatment may not be high, it has a disproportionate impact not only on public opinion, but on how governments and regulators respond. The risks to government arise because the actual prevalence of these events is not appreciated generally, and the appropriate response may not be clear because advice to governments from researchers, policymakers or industry on how to respond is lacking. The risks to other stakeholders lie in the threat to licence to operate as an animal production facility, whether an abattoir or a farm (Coleman, 2018; Martin & Shepheard, 2011). However, more routine human–animal interactions in livestock are particularly important because they have an impact on the welfare and productivity of farm animals even though they may not receive media scrutiny (Hemsworth & Coleman, 2011), and it is this that will be the main focus of this chapter. The people involved can be either farmers as owner-operators of the livestock farming facility or as employed stockpeople.

10.2 Impact of Human–Animal Interactions on Animal Welfare and Productivity

10.2.1 Context and Kinds of Interactions

The management of farm animals has undergone dramatic changes since the Second World War, especially in Western cultures such as European countries, the United States, Canada and Australia. Intensification of animal production from the 1950s involved a transition from traditional methods of keeping farm animals outdoors that relied on labour for routine tasks, such as feeding and manure removal, to systems in which animals are generally kept in specialised indoor environments using hardware and automation instead of labour for many routine tasks (Fraser, 2005). The impetus for this intensification was severalfold. After the Second World

DOI: 10.4324/9781003221753-10

War, Western governments developed policies to increase the availability of cheap and safe food, especially protein, for their populations (Cronin et al., 2014; Hodges, 2000). At the same time, producers had to increase production efficiency, such as animal growth and reproductivity efficiency while safeguarding animal health, to meet rising costs, which could be achieved through intensive animal production (Cronin et al., 2014; Hemsworth & Coleman, 2011). It has been proposed that intensification of animal production was not driven or directed by the voluntary choices of individual animal producers, but rather was determined by wider economic forces beyond the control of the producer (Fraser, 2005; Sandøe, 2008). Furthermore, while this change was occurring, animal production was becoming increasingly concentrated on fewer farms with increasing herd or flock sizes but a shrinking workforce. These pressures on animal production still exist today. Food security ranks high in global policy priorities; demand for food is increasing as populations grow and gain wealth to purchase more varied and resource-intensive diets (Garnett et al., 2013).

With global trends in increasing herd or flock sizes and decreasing workforce availability, there has been the emergence of new technologies to facilitate animal and facility monitoring, and to deliver time savings with recurrent physical tasks (e.g. milking, feeding) and increased flexibility in organising the work programme (Hostiou et al., 2017). Nevertheless, in modern intensive management of farm animals, human contact is at risk of becoming progressively more stressful as the number of animals managed by each stockperson increases, with many of the stockperson interactions biased towards negative ones, given that opportunities for positive human contact are probably minimal, and many routine husbandry tasks undertaken by stockpeople may contain negative elements, such as restraint, vaccinations and surgical interventions (Boissy et al., 2005; Hemsworth & Coleman, 2011). As discussed later, increased opportunity for stockperson interactions that are perceived as positive is likely to improve the human–animal relationship from the perspective of the animal, with implications for both animal welfare and productivity.

When managing farm animals in both extensive and intensive settings, and in relatively large groups but in restricted areas, stockpeople are able to interact regularly with their animals at several levels (Hemsworth & Coleman, 2011). Many of these interactions are associated with regular observation of the animals and their conditions. Animals in most production systems have to be moved and, in addition to visual and auditory contact, stockpeople often use tactile interactions to move sheep, goats, pigs and cattle, for example. At times, when animals are moved to unfamiliar or novel situations, stockperson interactions of a negative nature need to be used to move the animals. Some farm animals are rarely restrained during their lives, while others are restrained on a regular basis. Some sort of restraint is used for weighing, milking, vaccinating and blood sampling, and animals are restrained for procedures that are painful such as castration, branding, ear tagging and dehorning. It may be possible to reduce or eliminate some of these procedures or to minimise their effects by appropriate pain management. Procedures such as vaccinations and blood sampling for diagnosis are necessary to improve the health and thus the welfare of animals, and some degree of discomfort or pain is justified in the health and welfare interests of the animal. Procedures such as milking and shearing are directly related to the reason the animals are kept and could only be eliminated if the industry is shut down. Weighing, ear tagging, castration and dehorning are justified by facilitating management, improving product quality and/or reducing the possibility of injury to animals or humans. All these stockperson interactions with farm animals, therefore, contribute to the overall relationship that animals have with humans and determine whether the relationship is positive, neutral or negative. It is the nature of human interactions with the animal and the quality of the relationship from the animal's perspective that mainly determine the impact of human interactions on fear behaviour and the stress response of the animal.

10.2.2 Effects of Human Interactions on Fear Behaviour of the Animal

It is generally agreed that other animal species, particularly other mammals, are likely to have emotional or affective experiences that resemble our own (Panksepp, 2005), and most researchers studying human–farm animal relationships have predominantly focused on the fear responses of farm animals to humans because of their implications for animal productivity and welfare (Hemsworth & Coleman, 2011). Nevertheless, the animal's perception of the relationship is determined not only by negative emotional states, such as fear of humans, but also by positive emotional states, such as contentment, curiosity, play and companionship, generated by interaction with humans (Hemsworth & Coleman, 2011; Rault et al., 2020).

Different emotional experiences and motivations are likely to be involved in the animal's perception of and reaction to humans (Waiblinger et al., 2006), and these are likely to determine the strength of an animal's relationship with humans, which may vary from negative through neutral to positive. Furthermore, as proposed by Désiré et al. (2002), emotional experiences that are generated during interactions between humans and animals are likely to be determined not only by the properties of the other partner in the relationship but also by the perception of the other partner, and thus the interpretation of the whole situation. Emotional experiences, therefore, are likely to be elicited by a combination of basic evaluations, such as the suddenness, familiarity, valence, predictability and controllability of the interactions (see framework in Chapter 5, this volume). The consequences for the welfare of animals of emotional experiences arising from human interactions are obvious, particularly those human interactions leading to fear, and thus stress, in animals. These emotional experiences may also affect the ease of handling of the animal, as well as the safety of the handler, as, for example, fear may lead to defensive behaviour, such as aggression in large animals (Matos et al., 2015).

The most common behavioural tests used by scientists to assess fear of humans involve measuring the approach behaviour to a stationary human or the avoidance response to an approaching human (Hemsworth et al., 2018). These methods of measuring an animal's behavioural response to a human to assess fear are supported by the findings of behaviour and acute physiological stress response (i.e. glucocorticoid metabolite concentration) correlates in these tests (Hemsworth & Coleman, 2011).

Handling experiments, particularly with dairy cattle, pigs and poultry, have shown that negative or aversive handling, imposed briefly but regularly, will result in marked fear responses based on avoidance in the presence of humans (see reviews by Hemsworth & Coleman 2011; Hemsworth et al., 2018). Depending on the species, these negative handling treatments generally involved a brief slap, hit, shock with a battery-operated prodder, a shout and/or a fast speed or sudden movement whenever an animal approached or failed to avoid an approaching experimenter. These negative handling treatments utilised the types of stockperson interactions employed in the livestock industries and, although these treatments were imposed briefly over several weeks and only when the animals approached or failed to avoid the approaching experimenter, the negative handling treatments resulted in marked differences in fear of humans and stress relative to those animals handled positively, for example, with a pat, stroke, talking and/or slow predictable movement. Minimal handling when studied, generally resulted in moderate fear and no effects on sustained basal glucocorticoid concentrations (i.e. physiological stress levels).

10.2.3 Effects of Human Interactions on Farm Animal Stress Physiology, Productivity and Welfare

Surprisingly, while the treatments in these handling experiments described above were generally imposed regularly but briefly, these experiments showed that negative handling not only resulted in high fear levels but also adversely affected farm animal productivity in terms of growth, reproduction and health (see reviews by Hemsworth & Coleman 2011; Hemsworth et al., 2018). In general, field studies have also shown negative correlations between fear of humans and farm animal productivity (Table 10.1).

Physiological stress is implicated in these effects of fear on farm animal productivity, because the negative handling treatments that increased fear and reduced farm animal productivity also often resulted in sustained elevations in basal glucocorticoid concentrations. Fearful animals also showed an acute glucocorticoid response (i.e. increased physiological stress response) in the presence of humans. These handling experiments indicate that conditioned fear responses to humans develop as a consequence of associations between the handler and aversive and rewarding elements of the handling bouts, and thus farm animals consistently handled in a negative manner show high levels of fear of humans and acute stress in the presence of humans and are therefore at risk of chronic stress and reduced productivity.

There are also concerns about the welfare of animals that are highly fearful of humans. Fear is generally considered an undesirable emotional state of suffering in both humans and animals. The research described here shows that farm animals that are both highly fearful of humans and in regular contact with humans are likely to experience acute stress responses in the presence of humans as well as prolonged (chronic) stress responses. Furthermore, fearful animals are also more likely to sustain injuries in trying to avoid humans during routine inspections and handling (Hemsworth, 2019).

While high fear levels can adversely affect farm animal welfare and productivity, animals that perceive interacting with humans per se as rewarding, that is, have a positive relationship with humans, may receive distinct benefits in modern production systems (Hemsworth & Coleman, 2011; Rault et al. 2020). As discussed earlier, animals may experience positive or pleasant emotional experiences in the presence of humans that may arise from associating humans with rewarding events, such feeding, patting and grooming by humans, and show increased attraction to humans in terms of their approach behaviour (see Hemsworth, 2003). For example, stroking

Table 10.1 Correlations (inter-farm) between fear of humans and animal productivity in the livestock industries

Species	*Study*	*Productivity Variable*	*Correlations between Fear and Productivity*
Pig	Hemsworth et al. (1981)	Piglets/sow/year	−0.51*
	Hemsworth et al. (1989)	Piglets/sow/year	−0.55*
	Hemsworth et al. (1994a)	Piglets/sow/year	−0.01
	Hemsworth et al. (1999)	Stillborn piglets	0.44*
Dairy cow	Breuer et al. (2000)	Milk yield/cow	−0.46*
	Hemsworth et al. (2000)	Milk yield/cow	−0.27
Meat chicken	Hemsworth et al. (1994b)	Food to gain ratio	−0.57**
	Hemsworth et al. (1996)	Food to gain ratio	−0.49*
	Cransberg et al. (2000)	Mortality	−0.10
Laying hen	Barnett et al. (1992)	Egg production/hen	−0.58**

*, **: Significant correlations at $P < 0.05$ and < 0.01, respectively.

applied in a manner that is similar to intraspecific allogrooming has been shown to reduce heart rate and result in relaxed body postures and increased approach to humans in cattle (see reviews by Hemsworth & Coleman, 2011; Hemsworth et al., 2018; Rault et al. 2020). Animals that experience positive emotional experiences in the presence of humans may have reduced stress responses in unfamiliar or painful situations, such as veterinary inspections or husbandry interventions, or even when kept in suboptimal environments (see reviews by Hemsworth & Coleman, 2011; Hemsworth et al., 2018; Rault et al. 2020). For example, brief daily positive human contact has been shown to reduce the magnitude of the physiological stress response to tether housing of sows (Pedersen et al., 1998).

10.2.4 Effects of Human Interactions on Ease of Handling Farm Animals

While it may appear logical that animals that are more fearful of humans should be easier to move, these animals may be the most difficult to handle in unfamiliar locations (Hemsworth, 2019). Farm animals, like other animals, are generally wary of moving towards an unfamiliar or unpredictable situation, and if they are fearful of both this environment and the handler, then they are most likely to show exaggerated behavioural responses to handling, such as balking or fleeing back past the handler. For example, there is evidence that pigs that are highly fearful of humans are the most difficult pigs to move along an unfamiliar route; fearful pigs took longer to move, displayed more balking, and were subjectively scored as the most difficult to move by the handler (Hemsworth et al., 1994c). Furthermore, it is generally easier to impose management practices on animals that are less fearful of humans. Positive handling of dairy cows in the form of talking, feeding and stroking facilitated entry to the milking parlour and reduced restless behaviour such flinch, step and kick responses during milking (Breuer et al., 2000; Hemsworth et al., 2000; Waiblinger et al., 2002) and Waiblinger et al. (2004) found that previous positive handling of dairy cows reduced heart rates, kicking and restless behaviour in both the presence of humans and during the veterinary procedure of rectal palpation.

10.3 Human Factors that Influence Human–Animal Interactions

The human characteristics that motivate human behaviour towards farm animals under their care have been the subject of extensive research. Coleman (2004) identified a range of attributes that may characterise a good stockperson and, therefore, may contribute to positive stockperson animal interactions. These included a combination of dispositions (personality, empathy, attitudes) and learned factors (skills, knowledge). While it may be intuitively appealing to consider attributes such as personality or empathy as the key factors in determining the nature of human animal directions, these human factors generally do not give rise to specific behaviours. For example, the five-factor personality model (Costa & McCrae, 1992) can be useful in determining the fit between an individual and a particular job type, but it is not useful in predicting specific behaviours within that job (Judge et al., 2013). Accurate prediction of behaviour depends on the degree of specific match between the human factor on the one hand and the specific activity within a job on the other. Judge et al. (2013, p. 879) point out that the human traits used to predict behaviour need to be measured with 'the same level of generality (or specificity) as the behaviours they seek to predict'.

Coleman (2004) reviewed research that generally supported Judge et al.'s proposition. Seabrook (1972 a, b) reported that milk yield in dairy herds were higher in herds where stockpeople were introverted and confident, and where the cows were most willing to enter the milking shed and were less restless in the presence of the stockperson. The role of stockperson personality such as agreeableness and pessimism was investigated further by Waiblinger et al. (2002), who found that stockperson personal characteristics including personality or abilities, based on the measures used by Seabrook, did not correlate significantly with milk yield, but did correlate with the attitudes of stockpeople towards cows and towards working with cows. Attitudes were the best predictors of stockperson behaviour. This role of attitudes as the direct predictors of stockperson behaviour is consistent with the theory of planned behaviour (Ajzen, 1991). The theory of planned behaviour states that behaviour depends on both motivation (intention) and ability (behavioural control). It identifies three types of beliefs and their associated attitudes – behavioural, normative and control. Behavioural attitudes are those attitudes that reflect positive or negative evaluation of the behaviour of interest. Normative beliefs are beliefs about the extent to which others who are important to the person would approve or disapprove of the behaviour. Control beliefs refer to the perceived ability to perform the behaviour.

A pattern of results similar to those from the dairy industry has been found in pig research. As was the case for dairy herds, Seabrook (1996) reported that larger pig litter sizes were associated with stockpeople with confident personalities, emotional stability, independent personality, rational behaviour and low aggression. Ravel et al. (1996), using the Sixteen Personality Factor Questionnaire, a well-validated personality inventory, found that high stockperson insecurity and low sensitivity were associated with high piglet survival at independent owner-operated farms. They concluded that being overly self-assured or overly sensitive were detrimental to good management practices in the farrowing shed. They also found that stockpeople who were highly reserved and bold, suspicious, tense and less emotionally stable were associated with higher piglet mortality at large integrated farms.

These results by Seabrook and Ravel are characterised by farm outcomes as the dependent variables rather than individual stockperson behaviours and the impact of these behaviours on animal productivity and welfare. Many factors can intervene between stockperson characteristics such as personality and the productivity and welfare of the animals under their care.

One of the other main candidates for a human factor that may influence HAIs and their effects on welfare and productivity is empathy. Empathy has two basic components, a capacity to vicariously experience another's emotional state or a cognitive capacity for role-taking. While empathy is a dispositional characteristic, unlike personality, which is a stable human characteristic, there is some suggestion that it may be learned. In characterising empathy, Duan and Hill (1996) identified a trait approach in which empathy is proposed to be innate and a situation-specific social learning approach in which empathy is amenable to training. Only limited empirical data from agriculture on the relevance of empathy as a stockperson characteristic are available. Coleman et al. (1998) found that empathy towards animals was associated with positive beliefs about pigs and about handling pigs. In this study, empathy was not a direct predictor of behaviour, but was associated with positive attitudes towards working with pigs.

The most parsimonious theory to account for the human characteristics that are most directly relevant to the prediction of stockperson behaviour is that stockperson attitudes are the most important determinants of stockperson behaviour and that these attitudes are influenced by other stockperson background characteristics and personal traits. This approach is encapsulated in the theory of planned behaviour (Ajzen, 1991). We will briefly review the evidence in support of this approach.

Much of the research up until the early 2010s is summarised in Hemsworth and Coleman (2011). This research demonstrated that in the dairy, pig and poultry industries, stockperson attitudes towards the specific livestock species and, more particularly, towards working with their animals were significant predictors of their actual behaviour. Behaviour was defined specifically in terms of positive or negative interactions, where positive or negative behaviour was defined in objective terms – for example, in dairy cattle, slaps, hits and tail-twists while moving cows in and out of the milking facility were defined as negative behaviours, while the main positive behaviours were pats, strokes and a hand on the cow's flanks or legs during milking (Hemsworth & Coleman, 2011). Similar definitions were used for pigs. One of the key findings of this research was that positive stockperson behaviour was a good predictor of livestock production and that interventions to improve stockperson behaviour led to improved productivity and welfare (Hemsworth & Coleman, 2011). Andreasen et al. (2020) provided further indirect support for the relevance of stockperson attitudes in farm animal welfare by showing that positive farmer attitudes towards handling cows were associated with better overall animal welfare indicators based on the welfare quality indicators good feeding, good health, good housing and appropriate behaviour. Other recent research has provided further support for the role of attitudes and positive interactions in determining positive human–animal relationships. A recent study on dairy herds in Brazil (Ujita et al., 2020) found short-term improvements in behaviour, stress response and residual milk in Gyr heifers that had been exposed to positive handling prior to introduction to the milking experience. Lange et al. (2020) found that fear in unrestrained cows was reduced more following positive interactions than in restrained cows. Kauppinen et al. (2013) found that attitudes were related to humane treatment intentions in stockpeople, but they did not target intention in relation to specific stockperson behaviours nor did they assess actual stockperson behaviour. They found no relationships between intention and animal productivity, but it is hard to interpret these results because there were no data on intention–behaviour relationships. However, Ebinghaus et al. (2018) did find consistent positive relationships between stockperson attitudes and behaviour and cow behaviour as measured by avoidance behaviour. This study did not assess the relationship between stockperson behaviour and milk production.

Figure 10.1 summarises all of the forgoing research on the factors that motivate stockperson behaviour, the effects of stockperson behaviour on animal stress and fear responses and animal welfare outcomes as well as the limited data on the relationship between attitudes, behaviour and job satisfaction, and motivation to learn. The relationships between background factors,

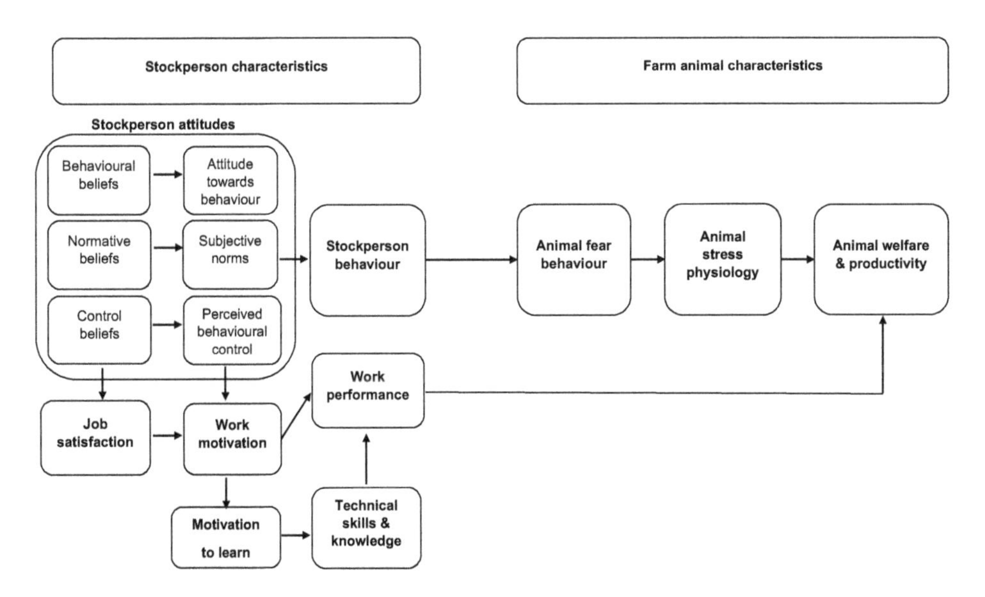

Figure 10.1 Model of stockperson–animal interactions with livestock adapted from Hemsworth and Coleman, 2011.

Figure 10.2 Prof. Coleman's talk on behavioural change for stockpeople in the livestock industries, Human Behaviour Change for Animals Conference, 2016. Scan the code to access the video (www.youtube.com/watch?v=C-Ixpwu6bdk&t=517s&ab_channel=HumanBehaviourChangeforAnimalsCIC).

attitudes and behaviour are based on Ajzen and Fishbein's theory of planned behaviour (Ajzen, 2005). The relationships between stockperson behaviour and animal fear and welfare are based on research by Hemsworth and Coleman (2011) and others reviewed earlier in this chapter. The other relationships between attitudes, job satisfaction motivation and technical skills are based on limited research by Hemsworth and Coleman (2011). While the research findings reviewed so far provide good evidence in support of this model, there are many gaps in the available data which will be addressed later in this chapter.

In summary, early findings and subsequent research indicate that stockperson interactions with the animals under their care have significant effects on farm animal welfare and productivity. This has led to interventions that aim to improve attitudes and behaviour of stockpeople, leading to an improvement in animal welfare and productivity. One example is the cognitive-behavioural programme called ProHand (Coleman & Hemsworth, 2014). This approach modifies stockperson attitudes and behaviour first, by providing information on the known relationships between fear, stress and productivity in farm animals, the effects of negative stockperson behaviour on animal fear and the way in which stockperson attitudes influence stockperson behaviour. This is then followed by feedback on the attitudes that lead to inappropriate stockperson behaviour, illustrated by examples and recommendations for behavioural change. Interventions using this approach have been shown to improve welfare and productivity in dairy cows and pigs (Hemsworth et al., 1994a; Coleman et al., 2000; Hemsworth et al., 2002). These studies not only demonstrated that stockperson attitudes and behaviour towards animals could be improved by appropriate training but also that a causal link exists between stockperson attitudes, behaviour, farm animal stress and productivity. These intervention studies provide evidence in support of the sequential model depicted in Figure 10.1 (see video for more information Figure 10.2).

10.4 Cost and Benefits of Human–Animal Interactions for Humans Including Mental Health

As reviewed earlier, stockperson interactions can have profound and sustained behavioural and physiological changes in farm animals, which can result in animal suffering and compromised animal productivity. While the benefits of improving farm animal welfare are difficult to evaluate from a purely economic perspective, it is widely recognised that animals with poor welfare are unlikely to produce at optimal levels (Hemsworth & Coleman, 2011). Furthermore, there appear to be other benefits of improving animal welfare that extend beyond production gains per se, since failure to safeguard animal welfare poses substantial risks to market protection, consumer acceptance and social licence to farm (Fernandes et al., 2021). While there are costs associated with training, the research on HAR indicates that stockperson attitudes are amenable to change and that cognitive behavioural training of stockpeople can improve animal welfare and productivity. Furthermore, farm animals that have a positive relationship with humans may receive distinct welfare benefits in modern production systems such as positive emotional experiences in the presence of humans and reduced stress responses in stressful situations (Rault et al., 2020).

There is some limited evidence that there is an association between stockperson attitudes and job satisfaction and stockperson well-being. For example, Coleman et al. (1998) showed significant correlations between negative beliefs about pigs, attitudes towards working with pigs, and beliefs about the characteristics of pigs on the one hand and low job enjoyment, anticipation of work breaks, and impact of work on family activity on the other. Maller et al. (2005) found negative relationships between farmer positive beliefs about cow behaviour and operator comfort while milking cows and work/life balance. More recently, Hansen and Østerås (2019) found that there was a positive relationship between Norwegian farmers' well-being (that had some attributes similar to those measured by Coleman et al. (1998)) and farm animal welfare. Farmer stress was negatively associated with animal welfare. While these results do not implicate effects on HAIs, given the relationships observed by Coleman et al. (1998), it is likely that farmer well-being and stress would impact HAIs. King et al. (2021) found that, amongst Canadian farmers, farmer stress and anxiety were associated with higher levels of severe lameness in dairy cattle. Rouha-Mulleder et al. (2009) found that, while features of the lying surface and free-stall design were the most important risk factors, the frequency of positive interactions by the stockperson, such as stroking and talking, was predictive of lameness; increased use of positive interactions was related to reduced prevalence of lameness. Also, while HAIs are not directly implicated, given the association between the prevalence of lameness and stockperson impatience when moving cows (Chesterton et al., 1989; Dewes, 1978; Eddy & Scott, 1980), there seems likely to be a relationship between farmer mental state, HAIs and welfare. All of the forgoing suggests that there would be a constellation of psychological states that would co-occur amongst stockpeople, their interactions with their animals and farm animal welfare, but these have not been systematically explored.

10.5 Conclusion

The importance of human–animal relationships in the livestock industries has been well demonstrated in a range of species and contexts. However, the distribution of relevant research is very uneven. Much of the work has been done in the context of on-farm activity in the pig and dairy industries. Limited research has been carried out in abattoirs (Coleman et al., 2012; Hemsworth et al., 2019), and only very limited work has been done in relation to animal transport. The causal relationships between stockperson characteristics, animal behaviour and animal stress responses and outcomes in terms of productivity on farms or in feedlots, or meat quality at abattoirs do, in principle, apply in all of these contexts. However, there must be specificity in relevant stockperson characteristics, animal responses and welfare and production outcomes. For example, relevant stockperson attitudes need to be determined in relation to the kinds of interactions that the stockperson has with the animal under their care. In the case of dairy cattle, these attitudes would comprise stockperson beliefs about the kinds of contact that are appropriate when regularly moving cows from pasture or barn to milking facility, and in and out of the milking facility, or beliefs about the amount of force or effort that is appropriate to be used in moving cows, and beliefs about desired speed of movement. In the case of pigs, however, while the relevant stockperson behaviours may be similar, the context in which these behaviours would occur would be regularly inspecting and moving breeding pigs and growing pigs to specialised accommodation at specific stages of reproduction or growth. Thus, for each species, attitude assessments need to be framed in such a way as to specifically relate to the particular species and context. When assessing animal behaviour and fear responses, the same principles apply. Whether assessing fear of humans using an approach test or a withdrawal test, the testing situation needs to be constructed to apply specifically to the species and the context in which the relevant behaviour occurs. It is equally obvious that the production outcomes are also species and context dependent. These outcomes may include such factors as milk yield, reproductive performance, growth rate or meat quality.

The main reason why these specific factors need to be assessed is that, in order to develop an intervention to change human attitudes and behaviour leading to improved human–animal relationships, the intervention needs to target those factors that are specifically relevant to the species and context.

Several approaches to modifying stockperson interactions with farm animals have been reported. Low Stress Stockhandling (LSS) (www.lss.net.au/index.htm) is designed to improve the stockperson–animal relationship and is based on the principles of livestock handling developed by Bud Williams (www.stockmanship.com). There is no experimental research that has evaluated the LSS approach, but there are numerous testimonials in support of it. LSS is not a behaviour change programme that systematically targets habitual behavioural patterns, the attitudes that underpin them, or the effects of peer pressure to conform to traditional handling methods and other factors that may make the stockperson resistant to change. Studies in the dairy and pig industries (Hemsworth et al., 1994a; Coleman et al., 2000; Hemsworth et al., 2002) have shown that cognitive-behavioural training, in which the specific attitudes and behaviour of stockpeople are targeted, can successfully improve the attitudes and behaviour of stockpeople towards their animals, with consequent beneficial effects on animal fear and productivity.

This research has underpinned the development training programmes to change stockperson attitudes and behaviour (ProHand: Coleman & Hemsworth, 2014; Quality Handling: Ruiz et al., 2010). At present, most versions of ProHand have targeted on-farm handling of livestock, although versions of ProHand have been developed for stockpersons in pig and red meat abattoirs. ProHand (Pig) has been implemented widely in Australia and New Zealand (over 2,000 stockpeople (ProHand (Pig)) and 300 abattoir stockpeople (ProHand (Pork Abattoir)) and several overseas organisations in the United States and Canada are using ProHand (Pigs). The main impediment to the implementation of these training programmes with other livestock species has been the lack of relevant basic and applied research.

There are many gaps in research to develop and evaluate training programmes in a range of areas. These include most kinds of extensive livestock farming, livestock transport both on land and on sea, and livestock in feedlots and in most poultry species. Such research needs to identify the stockperson-specific behaviours that impact on each species in each context, the impact of these behaviours on animal welfare and productivity, and attitudes that underpin stockperson behaviour.

As indicated earlier, there is little research that seeks to integrate human–animal relationships with stockperson and farmer quality of life, work satisfaction and work-life balance. In an environment where farming practices are under increasing scrutiny, where climate change and other environmental pressures increase stresses on farmers, it may be difficult for stockpeople to maintain a rational and sensitive approach to handling their animals. Research would aid in the development of comprehensive programmes that would target the entire work situation as well as best practices in handling the animals under their care.

10.6 Learning Outcomes

Source: Illustrated by Loïc Maréchal adapted from istockphotos.

Chapter Summary

- In modern intensive management of farm animals, human contact is at risk of becoming progressively more stressful as the number of animals managed by each stockperson increases.
- Many stockperson interactions are biased towards negative ones, given that many routine husbandry tasks undertaken by stockpeople may contain negative elements, such as restraint, vaccinations and surgical interventions.
- Negative handling not only results in high fear levels and, therefore welfare, but also adversely affects farm animal productivity in terms growth, reproduction and health.
- Animals that experience positive emotional experiences in the presence of humans may have reduced stress responses in unfamiliar or painful situations, such as veterinary inspections or husbandry interventions.
- Cognitive-behavioural training, in which the specific attitudes and behaviour of stockpeople are targeted, can successfully improve the attitudes and behaviour of stockpeople towards their animals, with consequent beneficial effects on animal welfare and productivity.

Check Your Understanding

Multiple Choice Questions

Question 1: Negative handling by stockpeople results in

- Answer a: High levels of fear of humans in farm animals but does not affect their growth or reproductive performance.
- Answer b: High levels of fear of humans and aggressive behaviour towards group mates in farm animals.
- Answer c: High levels of fear of humans, stress and reduced growth and reproductive performance in farm animals.

Question 2: Stockperson attitudes influence the quality of the human–animal relationship by determining:

- Answer a: how much training the stockperson requires.
- Answer b: the empathy of the stockperson.
- Answer c: the stockperson's behaviour.

Question 3: The best way to predict how stockpeople will interact with their animals is by knowing

- Answer a: their personality.
- Answer b: what their attitude is towards the activity itself.
- Answer c: their knowledge of animal behaviour.

Open Questions

1 Human–animal interactions may have as a comparable effect on the welfare and productivity of farm animals as the housing facility. Discuss.
2 To what extent does the well-being of farmers depend on positive human–livestock interactions, and how can we promote sustainable farming practices that prioritise the welfare of both farmers and animals?

11 Human–Wildlife Interaction

Laëtitia Maréchal, Tracie McKinney, Rie Usui and Catherine M. Hill

11.1 Introduction

The term 'wildlife' can be defined differently, and often specifically refers to free-living non-domesticated fauna, although it may also include flora. Depending on the context, such as different cultural perspectives, it might be defined, for example, as animals living free in their natural environment, or all non-owned animals, or as any living beings that are neither human nor domesticated (Décory, 2019). Indeed, these definitions can become unclear when wildlife interacts with humans. For instance, it might be debated whether the use of the term wildlife is appropriate for animal species living free in an urban environment but relying on human food such as waste, or captive animals that might have once lived free in their habitat. In this chapter, we define wildlife as any non-domesticated non-human animals, including captive 'wild' animals, but excluding companion and livestock animals, which are discussed in other chapters. We will explore the different aspects of human–wildlife interactions, including sharing space with wildlife and using wildlife for entertainment purposes.

11.2 Sharing Space with Wildlife

11.2.1 Cross-Cultural Perspectives

Western views of animals have been greatly influenced by Judeo-Christian beliefs and classical Greek thought, with a strong distinction made between what is animal and what is human. However, this stark separation between people and other animal species is not universally upheld. If we examine this phenomenon cross-culturally, we see that the distinction between what it is to be 'human' or 'animal' is culturally and historically dependent; it can shift; and the reasons for assigning people and animals to either side of this human–animal boundary can change. Animals may be considered clan members and part of people's kin groups, as totems and as ancestors or descendants of human ancestors. The Guajà people of Brazil hunt monkeys for food. Infant monkeys whose mothers are killed, are looked after by women and girls, who care for the young monkeys as if they were their own children. Female fertility is highly valued in Guajà society, and these infant primates are incorporated into their human caretaker's kin relations, thereby adding to a woman's fertility standing (Cormier, 2003). Animal totems are important figures to whom members of the clan are linked by ancestry and who are thought to provide protection to members of that clan. Clan members often abstain from harming or eating their totem animal. For example, among the Nuer of Sudan, members of the crocodile clans believed they were protected from crocodile threats provided they neither killed nor ate crocodiles (Evans-Pritchard, 1957 cited in Pooley, 2016). However, adherence to totemic beliefs

DOI: 10.4324/9781003221753-11

and practices can change over time, particularly among younger generations. Consequently, practices and beliefs documented previously are not necessarily upheld in more recent times.

Animals also feature in origin myths, as the ancestors of different human groups. Here they are linked to specific human groups through shared ancestors, or as individuals who were once human but, often because of a serious social transgression, were transformed from human into animal form as punishment for their misdemeanours. Origin myths from around the world centre on a variety of different wild animals, including wolves, crocodiles and primates. Among the Bari of Venezuela, there is the belief that the first spider monkeys were Bari people, who were transformed into spider monkeys by the creator of the Bari when they refused to share their food with other Bari people (Lizarralde, 2002). Animals can inhabit the spirit world as gods, or act as messengers between the spirit and human worlds. The Chinese mythical character Monkey King acts as a messenger between the gods of the spirit world and people on Earth (Burton, 2002). For the Ainu of Japan, the bear was the head of the gods and when the bear god visited the earth, it took on the form of a bear. Bears were hunted by the Ainu for food, and they also sometimes captured bear cubs that were nurtured and hosted by the Ainu until their release back to the spirit world through ritual killing as part of the bear ceremony (Yamada, 2018).

Ways of thinking about and understanding animals are linked to cultural, societal and experiential influences. Consequently animals, including wild animals, mean different things to different people. The wolf provides an excellent example of this. In many Western cultures, wolves have been reviled and feared for a very long time, and this antipathy towards wolves has persisted, even after they were eradicated from many parts of Europe, including Britain, centuries ago (Marvin, 2012). However, this animosity towards wolves is neither a geographical nor historical universal. Numerous Indigenous North American peoples have traditionally held wolves in high regard, in some cases voicing a strong feeling of identity with wolves (Marvin, 2012).

Certain wild animal species may also be viewed differently by members of the same society. Badgers in the United Kingdom represent an interesting conundrum. They are listed as Least Concern (IUCN Red List), are implicated in the transmission of bovine tuberculosis (bTB), and consequently are considered 'pests' by some groups, yet are legally protected from persecution, and are considered charismatic animals, worthy of protection. Cultural framings of badgers into 'good badger' and 'bad badger' largely match environmental and agricultural framings of the bTB issue and therefore, anti- and pro- badger cull sympathies (Cassidy, 2017). There are many other examples of contested species, where animals can either simultaneously occupy distinct, and often contradictory, categories, such as 'endangered' and 'pest', or wildlife species may shift from one category to another, as per white-tailed deer in the northeastern United States. Encounters between people and white-tailed deer have become increasingly common in suburban areas over time. When human–deer encounters were rare events, people put a high value on seeing the deer, but as encounters became more common and deer became more habituated to their human neighbours, the deer were no longer considered truly wild and their value declined (Leong, 2009). Leong (2009) refers to this as 'the tragedy of becoming common', and very likely this is relevant for other wildlife species which may well become less highly valued as a result of successful conservation initiatives that lead to increasing numbers in close proximity to human spaces.

11.2.2 Wildlife as Symbols

Wild animals represent many things to different people. For some, they are resources to be hunted, with weapons or cameras, their value often linked to their rarity, their supposed ferocity and/or the skills needed to track them down or overcome them. For others, wild animals are a staple or prized source of food, to be consumed within the household, or traded as a commodity to supplement household income. However, human–wildlife relationships are about much more than direct physical encounters or as a resource to be exploited for human benefit. Animals have symbolic meaning for people. Engaging with the symbolic nature of animals is a key aspect to understanding human–wildlife relationships more fully. Work by Skogen (2017) on people–wolf relationships in Norway identifies two different ways of thinking about wolves. Long-term rural residents associate the return of wolves to the Norwegian countryside with rural depopulation and declining quality of life in rural areas. For these people, wolves symbolise the devaluing of rural traditions, livelihoods and rural people. However, for middle-class, well-educated households who are moving into the Norwegian countryside in increasing numbers, the return of wolves is a symbol of authentic wilderness and the partial repair of the landscape after generations of anthropogenic change (Skogen, 2017).

There are many accounts of shapeshifting, or transformation from human to animal form, within the social sciences literature. Not all accounts of human transformation into animal form are associated with shapeshifters as agents of harm, but in many instances shapeshifted or were-animals (including European werewolves) are associated with ideas of people transforming into animal form to carry out aggressive and immoral acts against their fellow humans. For example, in southwestern Guinea-Bissau, people and chimpanzees are sympatric (share the same space) and very occasionally people experience attacks or threatening behaviour from wild chimpanzees. The circumstances in which these aggressive encounters occur are linked to the way local people understand the event. Attacks on people by 'clean' chimpanzees are viewed as a normal response by an animal defending its infant, itself or its access to a desired food resource. Attacks that appear unprovoked are understood to be by an 'unclean' or 'shapeshifted' chimpanzee (i.e. a sorcerer who has transformed into chimpanzee form to carry out harmful actions against other people). These 'unclean' events are symbolic of, and understood as, conflicts between relatives, and not as attacks by wild animals (Sousa et al., 2017).

11.2.3 Attitudes and Wildlife Values

Understanding people's attitudes towards animals, particularly certain wild animals, is often linked to ideas about how to manage or even change people's behaviour towards these species. Attitudes and behaviour are not necessarily strongly linked with each other. However, research into wildlife values and wildlife value orientations (WVOs), which characterise an individual's stable beliefs about wildlife, can provide information about people's attitudes and likely behavioural responses. Values are a person's underlying ideas about personal aspirations, such as trustworthiness, respect or loyalty for instance, and are culturally endorsed. A person's WVOs reflect their basic belief patterns about wildlife and give contextual meaning to the values people hold in relation to wildlife (Manfredo et al., 2009). Research suggests that people's attitudes are influenced by fundamental values, which are much less changeable than attitudes and only adjust slowly (Manfredo, 2008). Consequently, understanding more about the complex nature of attitudes, values, and WVOs, and how, when and why they might change could improve our understanding of, and predictions about, people's relationships with wildlife.

Early research using these concepts explored values and WVOs among different groups within American society. Results indicated that WVOs predicted attitudes towards wildlife. People who held a strong domination orientation (see Table 11.1) were much more likely to find management practices that cause direct harm to wildlife an acceptable option than were those who held a mutualist value orientation. More recent research shows that wildlife values in the western United States are shifting from domination to mutualism, and this is associated with

Table 11.1 Wildlife Value Orientation Characteristics (WVOC)

Domination/Utilitarian	Wildlife has practical value and should be managed for human benefit. People holding this WVO are more likely to express support for hunting and other ways of using wildlife, and more likely to consider management options that involve restraint or death of wildlife acceptable.
Mutualist	Humans and wildlife should live in harmony and wildlife is entitled to similar rights as humans. People holding this WVO are more likely to adopt behaviours to nurture wildlife, such as feeding, and are less likely to support activities that result in injury or death of wildlife.
Pluralist	Have both Utilitarian and Mutualist orientations – responses will be context specific.
Distanced	Less interested in wildlife, so values not strongly oriented towards wildlife or wildlife issues.

Source: Adapted from Dietsch et al., 2017 and https://sites.warnercnr.colostate.edu/wildlifevalues/home-page/what-are-value-orientations/.

changes in levels of income, education and urbanisation and their impacts on value orientations among the younger generations because of changing societal conditions (Manfredo et al., 2009).

WVO studies have been used in applied research to explore how different stakeholder groups value wildlife as a way of informing future conservation initiatives. Birds of prey such as hen harriers are legally protected in the United Kingdom, but are sometimes persecuted in areas where people shoot grouse because hen harriers predate grouse chicks. A study of WVOs among members of (i) pro-field sport organisations (field sports included were hunting, shooting and fishing), (ii) conservation groups, (iii) pro-raptor groups (groups specialising in the protection of birds of prey) and (iv) pro-bird groups (involved in bird protection, including the protection of birds of prey) revealed that people identified as holding mutualist wildlife values had more positive attitudes towards harriers, preferred less invasive hen harrier management strategies, and were less sympathetic towards grouse shooting than were participants who held utilitarian WVOs (St. John et al., 2019).

WVOs have also been used to examine the likely response to the reintroduction of previously extirpated species in Germany and the United States. German university students were surveyed to determine their support for the reintroduction of bison and wolves in Germany. Those with mutualistic WVOs were much more in favour of the species being reintroduced than were those who subscribed to a domination WVO (Hermann et al., 2013). In Colorado (USA), value orientations are strongly mutualistic in most parts of the state, and in line with this there is strong opposition to lethal control of wolves (Manfredo et al., 2021). A public ballot in November 2020 voted to bring back wolves to Colorado, which fits closely with what WVOs predict for this state.

11.2.4 When Human and Wildlife Interests Collide – Conflicts about Wildlife

Conflicts about wildlife, often called human–wildlife conflicts, are a particular type of human–wildlife interaction where wildlife presence and/or behaviours 'pose a direct and recurring threat to the livelihood or safety of people, leading to the persecution of that species and conflict about what should be done' (IUCN SSC Human–Wildlife Conflict & Coexistence Specialist Group). However, the terms we use to refer to people–wildlife encounters, including those labelled 'human–wildlife conflict', affects the way we understand and envisage these interactions (Peterson et al., 2010). By labelling particular negative encounters as 'human–wildlife conflict' this encourages people to conceptualise them as events of 'conscious antagonism between wildlife and humans' (Peterson et al., 2010, p. 75), putting emphasis on wildlife as 'perpetrators' and people as 'victims' of the animals' actions. Similarly, animals that cause damage by foraging in crops are often termed 'crop raiders' and their activities as 'crop raiding' (Humle & Hill, 2016). These overly simplistic framings have important implications for how we understand and respond to these events, making the human–animal dyad the focal or central component, whereas there is now extensive research available that illustrates that in many cases these 'human–wildlife conflicts' are a consequence of conflicts between different human groups, as a result of competing agendas, viewpoints and differential power relations, rather than as direct conflict between human and animal protagonists (Madden & McQuinn, 2014; Redpath et al., 2013). Debates about terminology continue within the conservation literature, with most recent developments focusing on ideas about shifting from a 'conflict' framing to one of 'human–wildlife coexistence' (Pooley et al., 2021). However, there is as of yet little consensus as to what 'human–wildlife coexistence' means, with definitions ranging from it being purely about sharing space with wildlife, or people and wildlife co-adapting to sharing space, to conflict and coexistence being opposite ends of a continuum, with human–wildlife coexistence being a conflict-free state (Hill, 2021).

11.2.4.1 Impacts of People and Wildlife on Each Other

The threats or challenges for people cohabiting with wildlife include direct impacts such as increased risks of injury, disease or death to people (Khumalo & Yung, 2013; Naughton-Treves, 1997). Furthermore, it puts human livelihoods and safety at risk through damage to crops and property (Hill, 2018), but also through predation on livestock and, occasionally, pets (Treves et al., 2004), or through competition over food and use of natural resources (e.g. overfishing; Cummings et al., 2019). There are additional, hidden, negative impacts of sharing space with wildlife, including impacts on mental health (e.g. fear or stress of being attacked by wild animals; Jadhav & Barua, 2012), but also the costs of lost opportunities (see example Box 11.1) associated with the time and resources spent guarding crops or livestock (Barua et al., 2013), impact on children's schooling (MacKenzie et al., 2015), gender inequality where women bear a disproportionate burden of these costs (Ogra, 2008), and impacts on livelihood security (Catapani et al., 2021; Hill, 2004).

Box 11.1 Lost Opportunity Costs: Protecting Crops and Livestock from Wildlife

People's responses to the challenges of sharing space with wildlife are variable. People may engage in active methods to protect crops or livestock, such as guarding (including herding), hunting and engaging in retribution killing, and/or passive methods such as erecting barriers and fences, including electric fences, traps, poisons, and visual, auditory or olfactory repellents. In some places guardian animals such as dogs, donkeys and alpacas (Marker et al., 2005) are used to protect domestic livestock against predation by carnivores. Both guarding and livestock herding require significant time and effort, and therefore this may conflict with other household tasks as well as opportunities to participate in paid off-farm work and to attend school (Hill, 2004; MacKenzie et al., 2015). These are what we refer to as 'lost opportunity costs'.

Sharing landscapes with humans, their livestock and crops also has impacts on wildlife. Positive aspects for wildlife include access to additional food sources (crops and livestock) that are spatially and temporally clumped, and highly predictable as to when they are available. For example, research into elephants that forage on crops suggests crop foraging can significantly improve foraging efficiency, with male elephants achieving a larger body size compared with their age mates who do not feed on crops (Chiyo et al., 2005). Large body size confers reproductive advantages on male elephants whereby they achieve musth at a younger age and maintain longer periods of musth; consequently, they have a longer reproductive lifespan than their smaller peers (Lee et al., 2011). Additionally, there is evidence that female elephants may prefer to mate with larger-bodied males (Chiyo et al., 2011), which would suggest crop foraging might be a beneficial strategy for male elephants. However, foraging on human foods is not without risk and animals who engage in these risky activities can, and do, incur costs, including increased exposure to disease, injury or death.

11.2.4.2 The Levels of Conflict Framework

Conflicts about wildlife are multifaceted and complicated, often involving different stakeholder groups whose beliefs, value systems, priorities and plans conflict (Madden, 2004; Redpath et al., 2013). The Levels of Conflict framework (Canadian Institute for Conflict Resolution [CICR], 2000) is a useful framework for understanding the complexities and nuances of conflicts related to wildlife. It has its origins in the field of peacebuilding but has been adapted by Madden and McQuinn (2014) for use with conflicts about wildlife. The Levels of Conflict framework (as illustrated in Figure 11.1) works from the premise that there are up to three levels of conflict that can occur within a conflict: dispute, underlying and identity-based conflict (CICR, 2000). Dispute-level conflicts relate to the evident, physical appearance

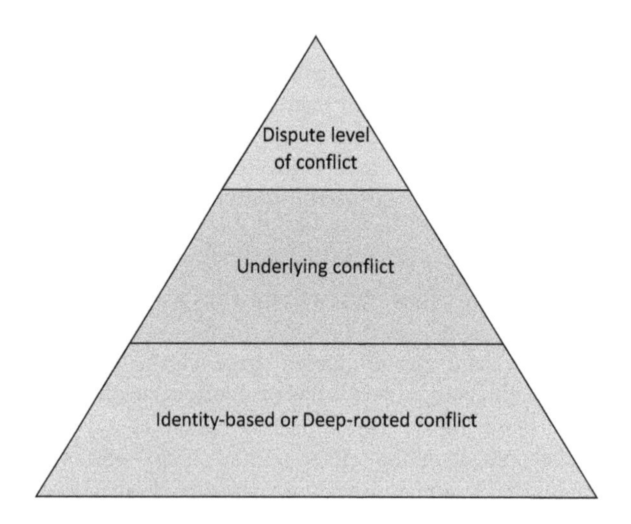

Figure 11.1 The 'Levels of Conflict' Framework.

Source: Adapted from Madden & McQuinn (2014).

Figure 11.2 Video talk by Prof. Alexandra Zimmerman, WildCRU Conservation Geopolitics Forum Plenary Talk (www.youtube.com/watch?v=DA37BfSmG-A&t=812s).

of the conflict (e.g. damage to a farmer's maize crop by foraging elephants). Underlying conflicts arise as a result of previously unresolved disputes that may, or may not, have anything to do with wildlife, but are a consequence of a previous grievance that went unnoticed or unanswered. This can include, for example, previous delay in or failure to receive monetary compensation for livestock losses to lion predation, or exclusion from decision-making processes about local land-use regulations. Identity-based or deep-rooted conflicts ensue when an individual or a group feels their sense of self is overlooked, or that they have been treated without respect and/or denied legitimacy in some way. Identity-based conflicts can be particularly difficult to describe or identify and are often articulated as material disputes in a conflict about wildlife, whereby complaints about the animal or its actions become a vehicle for expressing discontent about broader, deeper rooted, historical and identity-based conflicts (Madden & McQuinn, 2014). What we find is that, when technical solutions to reduce the impacts of wildlife on people's livelihoods and/or safety are implemented, irrespective of how effective, affordable or appropriate they may be, people may be very reluctant to implement them. Furthermore, even if they do adopt them, local conflict narratives may persist, irrespective of whether crop damage or livestock losses continue to occur (Dickman, 2010). You can find out more about the complex nature of conflicts about wildlife, and the need for transdisciplinary studies and understanding of these challenging scenarios in the video talk shown in Figure 11.2.

11.3 Human–Wildlife Interaction in Entertainment

11.3.1 The Different Wildlife Entertainment Contexts

11.3.1.1 Performing Wild Animals

11.3.1.1.1 CIRCUS, ZOO AND STREET PERFORMANCE

Animals used in performance date back thousands of years, with animals such as lions, elephants or bears used for Roman games (Nóbrega Alves & Duarte Barboza, 2018). Nowadays, performances using orcas, dolphins, elephants and non-human primates are very popular around the world. These species are used in venues ranging from zoos, theme parks, and circuses to small-scale shows. While circuses and zoos may claim that they are educating about wildlife, any animal performances must focus on species-normal behaviours and respect animal dignity to be effective educational activities (Brando, 2016). Numerous performances still include violent shows (e.g. animal fights), or the anthropomorphising of animals (e.g. monkeys performing human-like activities such as the dancing monkeys in Indonesia). However, cultural changes, animal welfare concerns, and legislation have led to increased debates about performing animals throughout the world. This has resulted, for instance, in a ban in the use of wild animals in circuses in several countries and states such as in Bolivia and some Mexican states. However, such events are currently permitted in some European countries and some states in United States.

11.3.1.1.2 WILDLIFE IN MEDIA/TELEVISION/FILM/ADVERT

Animals have featured in film and other entertainment media since at least the 1920s, as animal actors, animated characters and the subjects of wildlife documentaries (Nóbrega Alves & Duarte Barboza, 2018). The American Humane Association has been regulating the use of animals in film productions since 1939, although poor welfare and use of wild-caught animal actors persisted for many decades later (Molloy, 2011). Many domesticated animals have been used in cinema, but most welfare issues focus on the use of wild animals. Non-human primates, especially chimpanzees and capuchins, are among the best studied animal actors. While they appear infrequently in media (< 1% of films produced), the number of films depicting primate actors increased proportionately, with the number of films produced over the years 1990–2013 showing no decline in the use of these animals (Aldrich, 2018). Film producers historically have built a carefully constructed image of animal actors as autonomous individuals which enjoyed their role as 'stars', rather than as commodities (Molloy, 2011). As public knowledge and concerns for animal welfare have grown over the years, audiences today are more likely to view animal actors as exploited than as willing participants in the industry (Molloy, 2011). While many companies have pledged to no longer use apes in media, this restraint has not been extended to smaller primates (Leighty et al., 2015), and industry standards for the use of animals in film remain inconsistent and poorly regulated (Hitchens et al., 2021; Iacona, 2012).

The way animals are portrayed in film and other media can have a dramatic impact on popular perception. There are two conflicting viewpoints on animal representation in media. The 'familiarity' hypothesis argues that any depiction of wildlife makes them more familiar and likeable, and therefore encourages an interest in their conservation. The 'distortion' hypothesis argues that regular depiction of animals in media leads to the perception that they are not endangered in the wild, and perhaps would even make good pets. Studies have found strong support for the distortion hypothesis (Leighty et al., 2015; Schroepfer et al., 2011) but little support for the familiarity hypothesis. For example, viewing wild animals in advertisements has been shown to lead to the perception that those animals are not endangered (Brando, 2016; Leighty et al., 2015). Seeing animal behaviour out of context can also lead to misunderstandings. For example, bared-teeth expressions are common in primate actors in film but never in the context of fear or anger, when the expression would naturally be used (Aldrich, 2018). Likewise, viewers believed young chimpanzees in videos were adults, showing a fundamental misunderstanding of how large and powerful an adult chimpanzee would be (Schroepfer et al., 2011).

11.3.1.2 Wildlife in Captivity

11.3.1.2.1 ZOO AND WILDLIFE PARKS

Despite the global popularity of animal-based entertainment, the public is widely unaware of the differences between wildlife sanctuaries, zoos and more exploitative organisations (Doyle, 2017). We can define these different facilities by their aims and their dedication to animal welfare. Zoos and aquariums (hereafter 'zoos') are permanent institutions that maintain and display captive populations of wild animals. Zoos are accredited by regional bodies and the World Association of Zoos and Aquariums (WAZA), which ensure that welfare standards are met and encourage conservation and education as part of institutional mission statements (Association of Zoos and Aquariums, 2021; British and Irish Association of Zoos and Aquariums, 2021). Conservation is a main topic in zoo mission statements. Zoos often aim to function as an 'ark' for species, maintaining healthy populations until such time as conditions in the wild have stabilised (Bowkett, 2008). However, the majority of well-funded zoos are found in temperate regions, creating environmental mismatch with the animal's intended home range (Conde et al., 2011). The 'ark' model of species conservation is also constrained by taxonomic bias; zoos primarily house animals that will attract visitors, which may not necessarily be those in greatest need (Conde et al., 2011; Fa et al., 2014). Other approaches to conservation focus on funding in situ projects, or those based in the animal's natural surroundings. Maynard et al. (2019) found that zoos with more collaborative partners and those that act as project leaders have a greater conservation impact than institutions which provide funding alone. Some critics argue that education, conservation and research are admirable goals, but the success of these programmes is rarely evaluated (Maynard et al., 2019; Reiser, 2017).

11.3.1.2.2 SANCTUARIES OR RESCUE CENTRES

A sanctuary or rescue centre is a facility that provides lifetime care to wild animals, aiming at enhancing species-specific natural behaviours as much as possible, prioritising individual welfare over species conservation or welfare (Doyle, 2017). Animals are rescued or retired from circuses, zoos, illegal trades, laboratories or private owners. Many of these animals are kept in these facilities because they cannot be reintroduced into their natural habitat, due for example to health issues. In addition to their role for the animals in their care, such facilities have an opportunity to raise awareness about animal abuse, wildlife crime and conservation.

11.3.1.2.3 CONTROVERSY AND ETHICAL DEBATES ABOUT WILDLIFE CAPTIVITY

Understanding animal welfare needs for keeping animals in captivity is constantly improving. However, animal welfare standards are often slow to be implemented, and they differ between countries or institutions. For instance, posting pictures of visitors caring for infant primates used to be popular for rescue centres. However, such practices are now seen as inappropriate due to the negative impacts on primate welfare and conservation efforts (Waters et al., 2021). Furthermore, some organisations mislead the public by calling themselves sanctuaries, but they are actually linked to illegal or controversial animal exploitation practices, which are rarely associated with good animal welfare. For instance, some organisations in South Africa offering photo opportunities of cuddling lion cubs have been linked to unethical practices such as 'canned' hunting, which is killing highly habituated captive-bred wild animals in small enclosures (Hunter et al., 2013).

In addition to controversial practices in some captive organisations, keeping wild animals in captivity has raised numerous ethical debates (Doyle, 2017). For instance, animal life as experienced in their natural habitat will never be fully replicated in a captive environment. This includes inter- and intra-species interactions and spatial limitation. For example, captivity might reduce cognitive stimulations such as the ability to hunt for carnivores or complex multilevel social environments for elephants and chimpanzees. Animals in captivity are also constrained by human activities including those needed for their care (e.g. feeding, health checks, maintenance), as well as human choice (e.g. control over their reproduction). On the other hand, it is also argued that animals in captivity might have a safer and healthier life as they are not exposed to predation or human disturbance such as deforestation and poaching, and receive healthcare. Although there is no simple answer, these ethical debates are crucial in advancing our understanding and perception of wild animals in captivity, including animal exploitation.

11.3.1.3 Wildlife Tourism

Industrialisation and urbanisation caused a shift in human relationships with wildlife (Newsome et al., 2005). The industrialisation of society enhanced efficiency in production, which allowed people to enjoy more leisure time. Meanwhile, urbanisation separated people from nature and consequently generated a fascination with wildlife. People have increasingly participated in wildlife-related activities, such as wildlife viewing, which were formerly practiced exclusively by the elite (Franklin, 1999). This is known as 'wildlife tourism', which overlaps with nature-based tourism (Reynolds & Braithwaite, 2001). Because of its link with nature-based tourism, wildlife tourism is sometimes equated with ecotourism, which refers to 'responsible travel to natural areas that conserves the environment, sustains the well-being of the local people, and involves interpretation and education' (The International Ecotourism Society, 2015). However, wildlife tourism and ecotourism are not necessarily equivalent; while the former revolves around activities, the latter is based on values (Fennell, 2020). Thus, not all wildlife tourism is ecotourism.

Wildlife tourism, in a wide sense, includes consumptive and non-consumptive forms, and their outcomes differ largely. The former involves intentionally killing or physically harming animals for leisure such as sport hunting and fishing (Higginbottom, 2004). On the other hand, the latter is generally characterised as those activities in which animals are simply viewed by tourists (Higginbottom, 2004), although some interactions might involve touching and feeding animals. Wildlife tourism can also be classified into wildlife-dependent and wildlife-independent forms, based on tourists' intentions (Higginbottom, 2004). For example, activities that focus on wildlife such as whale watching and sport hunting are wildlife-dependent. Nara deer and Miyajima deer in Japan (see Usui & Funck, 2017) provide a good example of wildlife-independent tourism, as tourists do not necessarily visit these sites to see them, although the deer can be part of the reason why tourists visit.

Tourist-animal interfaces are now globally ubiquitous. Wildlife tourism has become a pivotal industry for the economies of many countries (World Tourism Organization and Guangdong Chimelong Group, 2020). This trend is particularly true in developing countries (WTOGCG, 2020). It is no surprise that, before the Covid-19 pandemic, the wildlife tourism market was dominated by African nations, which made up 88% of tour operators' annual revenue (UNWTO, 2015). The growth in the popularity of wildlife viewing can be associated with a combination of several factors, such as the public's environmental concerns, disconnection from nature, exposure to wildlife documentaries, the social movement towards conservation, improved access through transportation systems, and advancements in living standards (Newsome et al., 2005).

In the last decade, wildlife tourism has taken a new turn. Emergence of human–animal studies has led to renewed interest in wildlife tourism, namely, ethical concerns (Borges de Lima & Green, 2017), including 'proper treatment of animals, habitat and ecosystems, and local communities and business, as well as appropriate behavior of tourists' (Burns, 2017, p. 213). Unfortunately, the tourism industry does not adequately address or consider the ethical treatment of wildlife (Fennell, 2014) despite its link with a wide range of animals.

11.3.2 The Human Dimension of Wildlife in Entertainment

In the following sections we define 'who' are the humans in the human–wildlife interactions in entertainment, what their motivations, perceptions and attitudes are. We then investigate the cost-benefits of these interactions for humans, while taking into consideration the animal perspective.

11.3.2.1 Wild Encounter: Tourist/Visitor/Volunteer/Spectator

People seek out wildlife encounters – whether as tourists, visitors, volunteers or spectators – that offer varying levels of engagement with wildlife. Opportunities for interactions with wildlife in entertainment range widely from simply viewing and photographing wildlife to coming into close contact with wildlife by, for example, feeding and touching. These people are typically motivated by a desire to see wildlife – especially rare species – in close proximity without jeopardising their own safety (Higginbottom, 2004; Newsome et al., 2005). As popular wildlife entertainment expands, the type of tourist has shifted from specialist to generalist, and attracted people from different cultural backgrounds, which has resulted in the diversification of perceptions and attitudes towards wildlife. There is a need for humans to interact with nature, including wildlife (Wilson, 1984; see also Chapter 2, this volume). Wildlife encounters elicit emotional responses, such as a sense of wonderment, and offer spiritual, emotional and physical benefits (Curtin, 2009). On the other hand, potentially negative impacts include injuries (Pagel et al., 2020) and disease transmission (Jones-Engel et al., 2005). Such risks can be greater when feeding is involved (Orams, 2002).

11.3.2.2 Business of Care: Management/Staff/Keeper

Business personnel who support the management of human–wildlife encounters include park staff, rangers and zookeepers, among others. These stakeholders mediate tourist–wildlife encounters by leading guided tours or by monitoring both tourist and animal behaviours. They also incorporate educational opportunities through the interpretation of nature, which zoos and national parks use in guided tours and exhibits in visitor centres to inform guests about the significance of their experiences (Moscardo et al., 2004). These business operators are interested in increasing satisfaction among tourists as well as promoting wildlife and habitat conservation while also maximising economic benefits (Higginbottom, 2004). In natural settings, where encounters with wildlife are not guaranteed but are expected by tourists, the chances of spotting a rare species of wildlife during a guided tour is predicted to enhance tourists' satisfaction (Curtin, 2013). Moreover, in such a setting, where no physical barriers separate humans from wildlife, park rangers and staff have a critical role in protecting tourists from harm (Curtin, 2010). For instance, at macaque tourism sites, park rangers are required to mediate tourist–wildlife encounters to prevent physical contact (Orams, 1996). Park rangers also herd monkeys to the viewing areas to prepare them for viewing by tourists (Knight, 2010; Usui et al., 2014).

Facilitating positive experiences for tourists can encourage them to support conservation messages and raise awareness. However, it is important to note that facilitating tourist–wildlife interactions through feeding can be controversial as it affects animals' ecology and increases inter- and intra-species aggression, which results in injuries or poor health (Orams, 2002). If animals are fed, the encounter with such animals becomes less authentic as well (Knight, 2010).

11.3.2.3 Impacts on the Surrounding Community

Wildlife exploitation can range from benefiting to negatively affecting the surrounding communities (i.e. those not directly involved in the care of the animals or as visitors). For example, tourism can provide important revenue that can be invested for capacity development, such as the creation of jobs, investments in education, health, art or infrastructure (i.e. roads, buildings; Mutanga et al., 2020). However, tourism is also often associated with monetary leakage and corruption that deprive surrounding communities of important incomes (Sandbrook & Adams, 2012; Sekhar, 2003). Wildlife exploitation can also exacerbate disturbance for the surrounding communities, such as tourism overcrowding, risks for human safety (e.g. pathogen transmission or injuries), or crop damages around wildlife tourism sites (Burns, 2004). Therefore, these mixed outcomes often lead to mixed attitudes towards wildlife within the same community or even within the same person at different times.

11.3.3 The Future of Human–Wildlife Interaction in Entertainment

Human–wildlife interactions for entertainment purposes is rapidly growing. However, serious concerns have been raised for animal health, welfare and conservation as well as the negative consequences for humans as shown in this chapter. This has highlighted the need for careful assessment for a sustainable future for these interactions.

Recently, with the development of computer-generated imagery, real animals have been replaced by virtual animals in movies and advertisement. While this change improves animal welfare as no real animal is used, the public perception of these animals might be altered, as highlighted in this chapter (e.g. weakening conservation awareness/efforts and increasing illegal wildlife exploitation). Therefore, solutions are urgently needed to transform possible negative impacts that virtual wildlife has on real wildlife populations to positive assets (Courchamp et al., 2018). For example, paying a fee to conservation organisations for copyright use of a species for media/marketing purposes might generate income for conservation and enhance global public interest in preserving biodiversity (Courchamp et al., 2018).

In addition, better policies on social media could help reduce poor practices, improve animal welfare and reduce illegal wildlife trade. For instance, Instagram and Facebook explicitly ban the sale of endangered animals, and a reporting system has been implemented. However, it has been shown to be ineffective in many countries (Nijman et al., 2021), and rarely prevents animal abuse. Some guidelines on responsible portrayal of primates have been published by the International Union for Conservation of Nature (Waters et al., 2021) highlighting the issues related to the presence of humans in close proximity to primates.

Recently, growing debates around the ethics of animals used as attractions in street performances/circuses, zoos and wildlife tourism have led to legislation and guidelines aiming at developing sustainable human–wildlife interactions by improving animal welfare and human safety. For example, the Association of British Travel Agents (ABTA) published the best wildlife viewing practice guidelines for travel businesses and animal experience suppliers (ABTA, 2013). This has resulted in major international tourism companies such as TripAdvisor and Kuoni in 2017 pledging to follow these guidelines and banning some animal attractions from their tourist holiday services/bookings (e.g. elephant rides). Such guidelines have not yet been developed at an international level, reducing the impact of these initiatives. Unfortunately, tourists are often not aware of the impacts of such attractions on animal welfare (Moorhouse et al., 2016). Therefore, having an objective welfare and conservation assessment of each attraction provided by suppliers would empower tourists to make an informed decision about the use of animals in entertainment.

Finally, the following figures present the main take-home messages to ensure/facilitate responsible interactions with wildlife on social media (see Figure 11.3) and in person (see Figure 11.4):

Figure 11.3 Being responsible on social media! Report animal abuse when you see it by clicking on the dot symbol representing the different options to contact the social media provider.

Source: Prime Earth Education: www.prime-earth.org.

Figure 11.4 Being a responsible wildlife tourist.

Source: Illustration Grace Murray-Robertson adapted from Primates tourism guidelines K.S. Morrow.

11.4 Conclusion

This chapter highlights the diversity and complexity of human–wildlife interactions and the importance of understanding the different cultural perspectives, attitudes and values attributed to wild animals when sharing space with wildlife or using wild animals for entertainment. Cohabiting with wildlife might be constantly changing/evolving, ranging from positive to null to negative outcomes for both humans and animals. The experience of interacting or sharing space with wildlife is as much a result of human values and politics around wildlife as it is about animal actions and the consequences of those actions. Therefore, bringing social and natural science methods together to assess the different ecological and social factors influencing human–wildlife interactions is essential.

Wild animals have been used for entertainment purposes for thousands of years, but recently ethical concerns (e.g. welfare and conservation issues) have been raised about whether and how wild animals are used. While there is some progress worldwide on understanding animal welfare resulting in better practices in some entertainment industries (e.g. circus, zoos or film industry), these are highly variable between countries and/or organisations. Cultural diversity and the internationalisation of wildlife entertainment highlight the divergent perspectives towards wildlife and their uses (e.g. traditional animal performances and animal welfare standards). Therefore, the different perspectives must be taken into consideration to fuel the debate about using wild animals in entertainment, which has an important role in directing the future of human–wildlife interactions.

11.5 Learning Outcomes

Source: Illustrated by Loïc Maréchal adapted from istockphotos.

Chapter Summary

- Animals have symbolic significance for humans, and recognising this is key to understanding the human–wildlife relationship.
- Disputes around wildlife are complicated and stem from disagreements and lack of understanding and respect for other people's beliefs, values and needs.
- Using wild animals for entertainment is a highly debated topic, raising issues related to ethical, welfare and conservation concerns.
- The future of using wildlife in entertainment must take into consideration and mediate these concerns for more sustainable and responsible human–wildlife interactions.

Check Your Understanding

Multiple Choice Questions

Question 1: What does WVOC stand for?

- Answer a: Wildlife View Orientation Choice.
- Answer b: Wildlife Value Orientation Characteristic.
- Answer c: Wildlife View Oriented Culture.

Question 2: What are the main missions of a sanctuary?

- Answer a: encouraging conservation and education.
- Answer b: maintaining species welfare.
- Answer c: prioritising individual welfare over species conservation/welfare.

Question 3: What should be avoided when interacting with wildlife on social media?

- Answer a: I should report any animal abuse I see on social media.
- Answer b: I should not share selfies taken with wild animals.
- Answer c: I should share dressed-up wild animal images.

Open Questions

1. What factors should be taken into account when managing human–wildlife interactions, particularly when balancing the interests of human livelihoods, animal conservation and welfare?
2. Critically evaluate the use of wild animals for entertainment purposes. Please choose between wildlife in captivity and performing animals as topics.

Multiple Choice Questions: The Answers!

Chapter 2

Question 1: Answer a
Question 2: Answer b
Question 3: Answer b

Chapter 3

Question 1: Answer c
Question 2: Answer b
Question 3: Answer c

Chapter 4

Question 1: Answer a
Question 2: Answer a
Question 3: Answer a

Chapter 5

Question 1: Answer a
Question 2: Answer c
Question 3: Answer b

Chapter 6

Question 1: Answer c
Question 2: Answer c
Question 3: Answer b

Chapter 7

Question 1: Answer a
Question 2: Answer c
Question 3: Answer c

Chapter 8

Question 1: Answer b
Question 2: Answer c
Question 3: Answer b

Chapter 9

Question 1: Answer b
Question 2: Answer c
Question 3: Answer b

Chapter 10

Question 1: Answer c
Question 2: Answer c
Question 3: Answer a

Chapter 11

Question 1: Answer b
Question 2: Answer c
Question 3: Answer c

Going the Extra Mile ...

Figure A.1 QRcode of the textbook website (https://human-animal-interactions.blogs.lincoln.ac.uk/) where the reader can find some additional resources, including additional references, video presentations, links to useful websites and so on.

Complementary Reading Recommendations

DeMello, M. (2012). *Animals and society: An introduction to human–animal studies*. Columbia University Press.

Flynn, C. (2005). *Social creatures.* Lantern Press.

Fudge, E. (2007). *Animal*. Reaktion Press.

Herzog, H. (2010). *Some we love, some we hate, some we eat: Why it's so hard to think straight about animals*. HarperCollins.

Hosey, G., & Melfi, V. (2018). *Anthrozoology: Human–animal interactions in domesticated and wild animals.* Oxford University Press.

Kalof, L., & Fitzgerald, A. (2007). *The animals reader: The essential classic and contemporary writings* (2nd ed.). Routledge.

Knight, J. (2020). *Animals in person: Cultural perspectives on human–animal intimacies*. Routledge.

Other recommendations are available through our website!

Please note that we do not manage the websites associated with the QR code links provided in this book, and so some links might get deactivated! However, the reader will be able to find alternative resources on our website.

References

Abe, K., & Watanabe, D. (2011). Songbirds possess the spontaneous ability to discriminate syntactic rules. *Nature Neuroscience, 14*(8), 1067–1074.

Ainsworth, M.D., & Bell, S.M. (1970). Attachment, exploration, and separation: Illustrated by the behavior of one-year-olds in a strange situation. *Child Development, 41*(1), 49–67. https://doi.org/10.2307/1127388

Ainsworth, M.D.S., Blehar, M., Waters, E., & Wall, S. (1978). *Patterns of Attachment.* Erlbaum.

Ajzen, I. (1991). The theory of planned behaviour. *Organizational Behaviour and Human Decision Processes, 50*, 179–211.

Ajzen, I. (2005). *Attitudes, Personality and Behaviour*. Open University Press.

Akey, J.M., Ruhe, A.L., Akey, D.T. et al. (2010). Tracking footprints of artificial selection in the dog genome, *Proceedings of the National Academy of Sciences. U.S.A., 107*(3), 1160–1165.

Albuquerque, N., Guo, K., Wilkinson, A., Resende, B., & Mills, D. (2018). Mouth-licking by dogs as a response to emotional stimuli. *Behavioural Processes, 146*, 42–45. https://doi.org/ 10.1016/j.beproc.2017.11.006

Albuquerque, N., Guo, K., Wilkinson, A., Savalli, C., Otta, E., & Mills, D. (2016). Dogs recognize dog and human emotions. *Biology Letter, 12*, 20150883. https://doi.org/10.1098/ rsbl.2015.0883

Alders, R.G., & Pym, R.A.E. (2009). Village poultry: Still important to millions, eight thousand years after domestication. *World's Poultry Science Journal, 65*(2), 181–190.

Aldrich, B.C. (2018). The use of primate 'actors' in feature films 1990–2013. *Anthrozoös, 31*, 5–21.

Allan, A.T.L., Bailey, A.L., & Hill, R.A. (2020). Habituation is not neutral or equal: Individual differences in tolerance suggest an overlooked personality trait. *Science Advances, 6*(28), 1–16. https://doi.org/10.1126/sciadv.aaz0870

Allan, F.K. (2021). *A Landscaping Analysis of Working Equid Population Numbers in LMICs, with Policy Recommendations*. Brooke. (thebrooke.org)

Alleyne, E., Tilston, L., Parfitt, C., & Butcher, R. (2015). Adult-perpetrated animal abuse: Development of a proclivity scale. *Psychology, Crime, & Law, 21*, 570–588. https://doi.org/10.1080/1068316X.2014.999064

Alleyne, E., & Parfitt, C. (2018). Factors that distinguish aggression toward animals from other antisocial behaviors: Evidence from a community sample. *Aggressive Behavior, 44*(5), 481–490. https://doi.org/10.1002/ab.21768

Alleyne, E., & Parfitt, C. (2019). Adult-perpetrated animal abuse: A systematic literature review. *Trauma, Violence, & Abuse, 20*(3), 344–357. https://doi.org/10.1177/1524838017708785

Allsopp, N. (2012). *K9 Cops: Police Dogs of the World.* Big Sky Publishing.

Alshawawreh, L. (2018). Sheltering animals in refugee camps. *Forced Migration Review, 58*, 76–8.

Alvarez, A.S., Pagani, M., & Meucci, P. (2012). The clinical application of the biopsychosocial model in mental health: A research critique. *American Journal of Physical Medicine & Rehabilitation, 91*(13), S173–S180.

Amendt, J., Krettek, R., & Zehner, R. (2004). Forensic entomology. *Naturwissenschaften, 91*(2), 51–65.

American Psychiatric Association. (2013). Diagnostic and statistical manual of mental disorders. American Psychiatric Association. https://doi.org/10.1176/appi.books.9780890425596

American Society for the Prevention of Cruelty to Animals (ASPCA). (2021). Shelter intake and surrender. www.aspca.org/helping-people-pets/shelter-intake-and-surrender.

Amiot, C.E., & Bastian, B. (2015). Toward a psychology of human–animal relations. *Psychological Bulletin, 141*(1), 6.

Anderson, J.R., & Gallup, G.G. (2015). Mirror self-recognition: A review and critique of attempts to promote and engineer self-recognition in primates. *Primates, 56*, 317–326. https://doi.org/10.1007/s10329-015-0488-9

Anderson, K.L., & Olson, M.R. (2006). The value of a dog in a classroom of children with severe emotional disorders. *Anthrozoös, 19*(1), 35–49. https://doi.org/10.2752/089279306785593919

Anderson, S., & Meints, K. (2016). Brief report: The effects of equine-assisted activities on the social functioning in children and adolescents with autism spectrum disorder. *Journal of Autism and Developmental Disorders, 46*(10), 3344–3352. https://doi.org/10.1007/s10803-016-2869-3

Andrade, S.B., & Anneberg, I. (2014). Farmers under pressure: Analysis of the social conditions of cases of animal neglect. *Journal of Agricultural and Environmental Ethics, 27*, 103–126.

Andreasen, S.N., Sandøe, P., Waiblinger, S., & Forkman, B. (2020). Negative attitudes of Danish dairy farmers to their livestock correlates negatively with animal welfare. *Animal Welfare, 29*, 89–98.

Andrukonis, A., & Protopopova, A. (2020). Occupational health of animal shelter employees by live release rate, shelter type, and euthanasia-related decision. *Anthrozoös, 33*(1), 119–131. https://doi.org/10.1080/08927936.2020.1694316

Archer, J., & Winchester, G. (1994). Bereavement following death of a pet. *British Journal of Psychology, 85*, 259–271. https://doi.org/10.1111/j.2044-8295.1994.tb02522.x

Arendt, M. (2016). Diet adaptation in dog reflects spread of prehistoric agriculture. *Heredity, 117*, 301–306.

Arendt, M., Fall, T., Lindblad-Toh, K., & Axelsson, E. (2014). Amylase activity is associated with AMY2B copy numbers in dog: Implications for dog domestication, diet and diabetes. *Animal Genetics, 45*(5), 716–722.

Arluke, A., & Atema, K. (2017). Roaming dogs. In L. Kalof (Ed.), *The Oxford Handbook of Animal Studies* (pp. 113–134). Oxford University Press.

Arluke, A., Levin, J., Luke, C., & Ascione, F. (1999). The relationship of animal abuse to violence and other forms of antisocial behavior. *Journal of Interpersonal Violence, 14*, 963–975. https://doi.org/10.1177/088626099014009004

Arnold, K., & Zuberbühler, K. (2012). Call combinations in monkeys: Compositional or idiomatic expressions?. *Brain and Language, 120*(3), 303–309.

Ascione, F.R. (1993). Children who are cruel to animals: A review of research and implications for developmental psychopathology. *Anthrozoös, 6*, 226–247. https://doi.org/10.2752/089279393787002105

Ascione, F.R. (2001). *Animal Abuse and Youth Violence*. US Department of Justice, Office of Justice Programs, Office of Juvenile Justice and Delinquency Prevention.

Association of British Travel Agents. (2013). Global welfare guidance for animals in tourism. ABTA. www.abta.com/sites/default/files/media/document/uploads/Global%20Welfare%20Guidance%20for%20Animals%20in%20Tourism%202019%20version.pdf

Association of Zoos and Aquariums. (2021). www.aza.org/

Axelsson, E., Ratnakumar, A., Arendt, M.J., Maqbool, K., Webster, M.T., & Lindblad-Toh, K. (2013). The genomic signature of dog domestication reveals adaptation to a starch-rich diet. *Nature, 495*, 360–364.

Baciadonna, L., Briefer, E.F., & McElligott, A.G. (2020). Investigation of reward quality-related behaviour as a tool to assess emotions. *Applied Animal Behaviour Science, 225*, 104968.

Baker, K.C. (2004). Benefits of positive human interaction for socially housed chimpanzees. *Animal Welfare, 13*(2), 239–245.

Baker, S.B., Knisely, J.S., McCain, N.L., Schubert, C.M., & Pandurangi, A.K. (2010). Exploratory study of stress buffering response patterns from interaction with a therapy dog dot. *Anthrozoös, 23*(1), 79–91 https://doi.org/dx.doi.org/10.2752/175303710X12627079939341.

Balluerka, N., Muela, A., Amiano, N., & Caldentey, M.A. (2015). Promoting psychosocial adaptation of youths in residential care through animal-assisted psychotherapy. *Child Abuse and Neglect, 50*, 193–205.

Banks, M., & Banks, W. (2002). The effects of animal-assisted therapy on loneliness in an elderly population in long-term care facilities. *Journals of Gerontology: Series A, Biological Sciences and Medical Sciences*, M428-32, 57. https://doi.org/10.1093/gerona/57.7.M428.

Baragli, P., Scopa, C., Maglieri, V., & Palagi, E. (2021). If horses had toes: demonstrating mirror self recognition at group level in Equus caballus. *Animal Cognition, 24*(5), 1099–1108.

Baran, B.E., Rogelberg, S.G., & Clausen, T. (2016). Routinized killing of animals: Going beyond dirty work and prestige to understand the well-being of slaughterhouse workers. *Organization, 23*(3), 351–369. https://doi.org/10.1177/1350508416629456

Barber, A.L.A., Mills, D.S., Montealegre-Z, F., Ratcliffe, V.F., Guo, K., & Wilkinson, A. (2020a). Functional performance of the visual system in dogs and humans: A comparative perspective. *Comparative Cognition and Behavior Reviews, 15*, 1–44. https://doi.org/10.3819/CCBR.2020.150002

Barber, A.L.A., Müller, E.M., Randi, D., Müller, C.A., & Huber, L. (2017). Heart rate changes in pet and lab dogs as response to human facial expressions. *ARC Journal of Animal and Veterinary Sciences, 3*, 46–55. https://doi.org/10.20431/2455-2518.0302005

Barber, A.L.A., Randi, D., Müller, C.A., & Huber, L. (2016). The processing of human emotional faces by pet and lab dogs: Evidence for lateralization and experience effects. *PLOS ONE, 11*, e0152393. https://doi.org/10.1371/journal.pone.0152393

Barber, A.L.A., Wilkinson, A., Montealegre-Z, F., Ratcliffe, V.F., Guo, K., & Mills, D.S. (2020b). A comparison of hearing and auditory functioning between dogs and humans. *Comparative Cognition and Behavior Reviews, 15*, 45–94. https://doi.org/10.3819/ CCBR.2020.150007

Barcelos, A.M., Kargas, N., Maltby, J., Hall, S., & Mills, D.S. (2020). A framework for understanding how activities associated with dog ownership relate to human well-being. *Scientific Reports, 10*(1), 11363. https://doi.org/10.1038/s41598-020-68446-9

Barnett, J.L., Hemsworth, P.H., & Newman, E.A. (1992). Fear of humans and its relationships with productivity in laying hens at commercial farms. *British Poultry Science, 33*, 699–710.

Baron-Cohen, S., Tager-Flusberg, H., & Lombardo, M. (Eds.). (2013). *Understanding Other Minds: Perspectives from Developmental Social Neuroscience*. Oxford University Press.

Barua, M., Bhagwat, S.A., & Jadhav, S. (2013). The hidden dimensions of human-wildlife conflict: Health impacts, opportunity and transaction costs. *Biological Conservation, 157*, 309–316.

Bass, M.M., Duchowny, C.A., & Llabre, M.M. (2009). The effect of therapeutic horseback riding on social functioning in children with autism. *Journal of Autism and Developmental Disorders, 39*, 1261–1267. https://doi.org/10.1007/s10803-009-0734-3

Bassette, L.A., & Taber-Doughty, T. (2013, June). The effects of a dog reading visitation program on academic engagement behavior in three elementary students with emotional and behavioral disabilities: A single case design. In Child & Youth Care Forum, *42*, 239–256, Springer US.

Bastian, B., Loughnan, S., Haslam, N., & Radke, H.R.M. (2012). Don't mind meat? The denial of mind to animals used for human consumption. *Personality and Social Psychology Bulletin, 38*(2), 247–256.

Batt, S. (2009). Human attitudes towards animals in relation to species similarity to humans: A multivariate approach. *Bioscience Horizons: The International Journal of Student Research, 2*(2), 180–190, https://doi.org/10.1093/biohorizons/hzp021

Bauman, Z. (2008). *The Art of Life*. Polity Press.

Bauman, Z. (2009). *Does Ethics have a Chance in a World of Consumers?* Harvard University Press.

Beauchamp, T.L. (2011). Rights theory and animal rights. In T.L. Beauchamp & R.G. Frey (Eds.), *The Oxford Handbook of Animal Ethics* (pp. 198–222). Oxford University Press.

Beauchamp, T.L., & Childress, J.F. (1979). *Principles of Biomedical Ethics*. Oxford University Press.

Beck, A., & Katcher, A. (1984). A new look at pet-facilitated therapy. *Journal of the American Veterinary Medical Association, 184*, 414–421.

Beck, A.M. (2014). The biology of the human–animal bond. *Animal Frontiers, 4*(3), 32–36.

Beckers, G.J., Berwick, R.C., Okanoya, K., & Bolhuis, J.J. (2017). What do animals learn in artificial grammar studies? *Neuroscience & Biobehavioral Reviews, 81*, 238–246.

Beckers, G.J.L., Bolhuis, J.J., Okanoya, K., & Berwick, R.C. (2012). Birdsong neurolinguistics: Songbird context-free grammar claim is premature. *Neuroreport, 23*, 139–145. http://dx.doi.org/10.1097/WNR.0b013e32834f1765

Beetz, A. (2002). *Love, Violence, and Sexuality in Relationships Between Humans and Animals*. Shaker.

Beetz, A. (2013). Socio-emotional correlates of a school-dog-teacher team in the classroom. *Frontiers in Psychology, 4*, 886.

Beetz, A.M. (2005a). New insights into bestiality and zoophilia. *Anthrozoos-Journal of the International Society for Anthrozoology, 18*, 98–119.

Beetz, A.M. (2005b). Bestiality and zoophilia: Associations with violence and sex offending. In A.L. Podberscek & A.M. Beetz (Eds.). *Bestiality and Zoophilia: Sexual Relations with Animals* (pp. 46–70). Purdue University Press.

Beetz, A., Julius, H., Turner, D., & Kotrschal, K. (2012). Effects of social support by a dog on stress modulation in male children with insecure attachment. *Frontiers in Psychology, 3*, 352. https://doi.org/10.3389/fpsyg.2012.00352

Beetz, A., Kotrschal, K., Hediger, K., Turner, D., Uvnäs-Moberg, K., & Julius, H. (2011). The effect of a real dog, toy dog and friendly person on insecurely attached children during a stressful task: An exploratory study. *Anthrozoös, 24*(4), 349–368.

Beetz, A., & McCardle, P. (2017). Does reading to a dog affect reading skills? In A. Fine & N. Gee (Eds.), *How Animals Help Students Learn: Research and Practice for Educators and Mental-Health Professionals* (pp. 12–26). Routledge.

Beirne, P. (2000). Rethinking bestiality: Towards a concept of interspecies sexual assault. In A.L. Podberscek, E.S. Paul, & J.A. Serpell (Eds.), *Companion Animals and Us: Exploring the Relationships Between People and Pets* (pp. 313–331). Routledge.

Ben-Aderet, T., Gallego-Abenza, M., Reby, D., & Mathevon, N. (2017). Dog-directed speech: Why do we use it and do dogs pay attention to it? *Proceedings of the Royal Society B: Biological Sciences, 284*(1846), 20162429.

Bennett, P.C., Trigg, J.L., Godber, T., & Brown, C. (2015). An experience sampling approach to investigating associations between pet presence and indicators of psychological well-being and mood in older Australians. *Anthrozoös, 28*(3), 403–420. https://doi.org/10.1080/08927936.2015.1052266

Benton, T. (1998). Rights and justice on a shared planet: More rights or new relations? *Theoretical Criminology, 2*(2), 149–75.

Bermejo, M., & Omedes, A. (1999). Preliminary vocal repertoire and vocal communication of wild bonobos (*Pan paniscus*) at Lilungu (Democratic Republic of Congo). *Folia Primatologica, 70*(6), 328–357.

Berns, G.S., Brooks, A.M., & Spivak, M. (2012). Functional MRI in awake unrestrained dogs. *PLOS ONE, 7*, e38027.

Bertenshaw, C., & Rowlinson, P. (2009). Exploring stock managers' perceptions of the human-animal relationship on dairy farms and an association with milk production. *Anthrozoös, 22*(1), 59–69. https://doi.org/10.2752/175303708X390473

Birch, J. (2017). Animal sentience and the precautionary principle. *Animal Sentience, 2*(16), 1.

Birch, J., Burn, C., Schnell, A., Browning, H., & Crump, A. (2021, November). *Review of the Evidence of Sentience in Cephalopod Molluscs and Decapod Crustaceans*. The London School of Economics and Political Science.

Bloom, P. (2001). Précis of how children learn the meanings of words. *Behavioral and Brain Sciences, 24*(6), 1095–1103.

Bloom, P. (2004). Can a dog learn a word? *Science, 304*(5677), 1605–1606.

Blouin, D.D. (2012). Understanding relations between people and their pets. *Sociology Compass, 6*(11), 856–869.

Blouin, D.D. (2013). Are dogs children, companions, or just animals? Understanding variations in people's orientations toward animals. *Anthrozoös, 26*(2), 279–294.

Bock, B.B., Van Huik, M.M., Prutzer, M., Eveillard, F.K., & Dockes, A. (2007). Farmers' relationship with different animals: The importance of getting close to the animals – case studies of French, Swedish and Dutch cattle, pig and poultry farmers. *International Journal of Sociology of Agriculture and Food, 15*(3), 108–125.

Boissy, A., Bouix, J., Orgeur, P., Poindron, P., Bibé, B., & Le Neindre, P. (2005). Genetic analysis of emotional reactivity in sheep: Effects of the genotypes of the lambs and of their dams. *Genetics Selection Evolution, 37*, 381–401.

Bonas, S., McNicholas, J., & Collis, G. (2000). Pets in the network of family relationships: An empirical study. In A.L. Poderscek, E.S. Paul, & A.J. Serpell (Eds.), *Companion Animals and Us: Exploring the Relationships Between People and Pets* (pp. 209–234). Cambridge University Press.

Bond, J., & Mkutu, K. (2018). Exploring the hidden costs of human–wildlife conflict in northern Kenya. *African Studies Review, 61*(1), 33–54.

Borges de Lima, I., & Green, R.J. (2017). Introduction: Wildlife tourism management and phenomena: A web of complex conceptual, theoretical and practical issues. In I. Borges de Lima & R.J. Green (Eds.), *Wildlife Tourism, Environmental Learning and Ethical Encounters* (pp. 1–17). Geoheritage, Geoparks and Geotourism. Springer, Cham. https://doi.org/10.1007/978-3-319-55574-4_1

Borgi, M., & Cirulli, F. (2015). Attitudes toward animals among kindergarten children: Species preferences. *Anthrozoös, 28*(1), 45–59.

Borgi, M., Cogliati-Dezza, I., Brelsford, V., Meints, K., & Cirulli, F. (2014). Baby schema in human and animal faces induces cuteness perception and gaze allocation in children. *Frontiers in Psychology, 5*, 411.

Borgi, M., Loliva, D., Cerino, S., Chiarotti, F., Venerosi, A., Bramini, M. et al., (2016). Effectiveness of a standardised equine-assisted therapy program for children with autism spectrum disorder. *Journal of Autism and Developmental Disorders, 46*(1), 1–9. https://doi.org/10.1007/s10803-015-2530-6

Bouchard, K.N., Dawson, S.J., & Lalumière, M.L. (2017). The effects of sex drive and paraphilic interests on paraphilic behaviours in a nonclinical sample of men and women. *Canadian Journal of Human Sexuality, 26*(2), 97–111. https://doi.org/10.3138/cjhs.262-a8

Bowkett, A.E. (2008). Recent captive-breeding proposals and the return of the ark concept to global species conservation. *Conservation Biology, 23*, 773–776.

Bowlby, J. (1969). *Attachment and Loss* (Vol. 1: Attachment). Basic Books.

Brady, H. (2018, 3 July). South Korea rules killing dogs for meat illegal, but fight continues. *National Geographic*. www.nationalgeographic.com/animals/article/south-korea-dog-meat-ban-animals

Brando, S. (2016). Wild animals in entertainment. In B. Bovenkerk & J. Keulartz (Eds.), *Animal Ethics in the Age of Humans* (pp. 295–318). Springer.

Brandt, M.J., & Reyna, C. (2011). The chain of being: A hierarchy of morality. *Perspectives on Psychological Science, 6*(5), 428–446.

Brelsford, V.L., Dimolareva, M., Gee, N.R., & Meints, K. (2020). Best practice standards in animal-assisted interventions: How the LEAD Risk assessment tool can help. *Animals, 10*(6), 974. https://doi.org/10.3390/ani10060974

Brelsford, V.L., Dimolareva, M., Rowan, E., Gee, N.R., & Meints, K. (2022). Can dog-assisted and relaxation interventions boost spatial ability in children with and without special educational needs? A longitudinal, randomized controlled trial. *Frontiers in Pediatrics, 10*.

Brelsford, V.L., Meints, K., Gee, N.R., & Pfeffer, K. (2017). Animal-assisted interventions in the classroom- A systematic review. *International Journal of Environmental Research and Public Health, 14*(7), 669. https://doi.org/10.3390/ijerph14070669

Bremhorst, A., Sutter, N.A., Würbel, H., Mills, D.S., & Riemer, S. (2019). Differences in facial expressions during positive anticipation and frustration in dogs awaiting a reward. *Scientific Reports, 9*(1), 1–13.

Breuer, K., Hemsworth, P.H., Barnett, J.L., Matthews, L.R., & Coleman, G.J. (2000). Behavioural response to humans and the productivity of commercial dairy cows. *Applied Animal Behaviour Science, 66*, 273–288.

British and Irish Association of Zoos and Aquariums. (2021). https://biaza.org.uk/

Brooke. (2016, 19 October). United Nations recognises role of working animals in food security. www.thebrooke.org/news/un-recognises-working-animals-role

Brooke. (2019a). Invisible livestock. www.thebrooke.org/sites/default/files/Advocacy-and-policy/Invisi ble-livestock-2019.pdf

Brooke. (2019b). Working together to transform the brick kiln industry. hwww.thebrooke.org/sites/default/files/Professionals/Brooke_Brick_Collaborative_Brief.pdf

Brooke. (2020). Brooke ensures rural communities in Afghanistan receive vital support amid COVID-19 pandemic. www.thebrooke.org/news/brooke-ensures-rural-communities-afghanistan-receive-vital-supp ort-amid-covid-19-pandemic

Brooke East Africa. (2016). *Turkana Donkey Welfare Project*. BEA and APAD. YouTube. www.youtube.com/watch?v=yhwJktxh2CQ

Brooke West Africa. (2021). *Socioeconomic Contributions of Donkeys in Burkina Faso*. Brooke. /www.thebrooke.org/research-evidence/socioeconomic-contribution-donkeys-burkina-faso

Brooks, H.L., Rushton, K., Lovell, K., Bee, P., Walker, L., Grant, L. et al. (2018). The power of support from companion animals for people living with mental health problems: A systematic review and narrative synthesis of the evidence. *BMC Psychiatry, 18*(1), 1–12.

Broom, D.M. (2014). *Sentience and Animal Welfare*. CABI.

Bruce, V., & Young, A.W. (2012). *Face Perception*. Psychology Press.

Brusini, I., Carneiro, M., Wang, C., Rubin, C.J., Ring, H., Afonso, S. et al. (2018). Changes in brain architecture are consistent with altered fear processing in domestic rabbits. *Proceedings of the National Academy of Sciences of the United States of Americ., 115*(28), 7380–7385.

Budianski, S. (1992). *The Covenant of the Wild*. Weidenfeld & Nicolson

Bufill, E., Agustí, J., & Blesa, R. (2011). Human neoteny revisited: The case of synaptic plasticity. *American Journal of Human Biology, 23*(6), 729–739.

Burgdorf, J., & Panksepp, J. (2001). Tickling induces reward in adolescent rats. *Physiology & Behavior, 72*(1), 167–173. https://doi.org/10.1016/S0031-9384(00)00411-X

Burnard, C.L., Pitchford, W.S., Hocking Edwards, J.E., & Hazel, S.J. (2015). Facilities, breed and experience affect ease of sheep handling: The livestock transporter's perspective. *Animal, 9*(8), 1379–1385. https://doi.org/10.1017/S1751731115000543

Burns, G.L. (2004). The host community and wildlife tourism. *Wildlife Tourism: Impacts, Management and Planning* (pp. 125–144). Common Ground Publishing.

Burns, G.L. (2017). Ethics and responsibility in wildlife tourism: Lessons from compassionate conservation in the Anthropocene. In I.B. Lima & R.J. Green (Eds.), *Wildlife Tourism, Environmental Learning and Ethical Encounters* (pp. 213–220). Geoheritage, Geoparks and Geotourism. Springer, Cham.

Burton, F.D. (2002). Monkey king in China: Basis for a conservation policy? In A. Fuentes & L.D. Wolfe (Eds.), *Primates Face to Face: The Conservation Implications of Human-Nonhuman Primate Interconnections* (pp. 137–162). CUP.

Burton, L.E., Qeadan, F., & Burge, M.R. (2018). Efficacy of equine-assisted psychotherapy in veterans with posttraumatic stress disorder. *Journal of Integrated Medicine, 1*, 14–19. https://doi.org/10.1016/j.joim.2018.11.001

Buschdorf, J.P., & Meaney, M.J. (2015). Epigenetics/programming in the HPA axis. *Comprehensive Physiology, 6*(1), 87–110.

Buttelmann, D., & Tomasello, M. (2013). Can domestic dogs (*Canis familiaris*) use referential emotional expressions to locate hidden food? *Animal Cognition, 16*, 137–145. https://doi.org/10.1007/s10 071-012-0560-4

Butterfield, M.E., Hill, S.E., & Lord, C.G. (2012). Mangy mutt or furry friend? Anthropomorphism promotes animal welfare. *Journal of Experimental Social Psychology, 48*(4), 957–960.

Buttner, A.P., Thompson, B., Strasser, R., & Santo, J. (2015). Evidence for a synchronization of hormonal states between humans and dogs during competition. *Physiology & Behavior, 147*, 54–62. https://doi.org/10.1016/J.PHYSBEH.2015.04.010

Byrne, R.W., Cartmill, E., Genty, E., Graham, K.E., Hobaiter, C., & Tanner, J. (2017). Great ape gestures: Intentional communication with a rich set of innate signals. *Animal Cognition, 20*(4), 755–769.

Caeiro, C., Guo, K., & Mills, D. (2017). Dogs and humans respond to emotionally competent stimuli by producing different facial actions. *Scientific Reports, 7*(1), 1–11.

Callicott, J.B. (1980). Animal liberation: A triangular affair. *Environmental Ethics, 2*, 311–338.

Campbell, M., & MacKay, K.J. (2009). Communicating the role of hunting for wildlife management. *Human Dimensions of Wildlife, 14*(1), 21–36, https://doi.org/10.1080/10871200802545781

Canadian Institute for Conflict Resolution. (2000). *Becoming a Third-Party Neutral: Resource guide*. Ridgewood Foundation for Community-Based Conflict Resolution.

Caporael, L.R., & Heyes, C.M. (1997). Why anthropomorphize? Folk psychology and other stories. In R. Mitchell, Nicholas S. Thompson, & H.L. Miles (Eds.), *Anthropomorphism, Anecdotes, and Animals* (pp. 59–73). State University of New York Press.

Capponi, G. (2020). Candomblé rituals and food practices in Italy and Brazil: An ethnographic comparison. *Antropolitica – Revista Contemporanea de Antropologia, 48*(1), 119–141. https://doi.org/10.22409/antropolitica2020.0i48.a42106

Carder, G., Ingasia, O., Ngenoh, E., Theuri, S., Rono, D., & Langat, P. (2019). The emerging trade in donkey hide: An opportunity or a threat for communities in Kenya? *Agricultural Sciences, 10*(9), 1152–1177.

Carlisle, G.K. (2014). Pet dog ownership decisions for parents of children with autism spectrum disorder. *Journal of Pediatric Nursing, 29*(2), 114–123. https://doi.org/10.1016/j.pedn.2013.09.005

Carlstead, K. (2009). A comparative approach to the study of keeper-animal relationships in the zoo. *Zoo Biology, 28*(6), 589–608. https://doi.org/10.1002/zoo.20289

Carr, D., Friedmann, E., Gee, N.R., Gilchrist, C., Sachs-Ericsson, N., & Koodaly, L. (2021). Dog walking and the social impact of the COVID-19 pandemic on loneliness in older adults. *Animals, 11*(7), 1852. https://doi.org/10.3390/ani11071852

Carr, S., & Rockett, B. (2017). Fostering secure attachment: Experiences of animal companions in the foster home. *Attachment & Human Development, 19*(3), 259–277.

Cassels, M.T., White, N., Gee, N., & Hughes, C. (2017). One of the family? Measuring young adolescents' relationships with pets and siblings. *Journal of Applied Developmental Psychology, 49*, 12–20.

Cassidy, A. (2017). Badger-human conflict: An over looked historical context for bovine TB debates in the UK. In C.M. Hill, A.D. Webber, & N.E.C. Priston (Eds.), *Understanding Conflicts About Wildlife. A Biosocial Approach* (pp. 65–94). Berghahn Books.

Catapani, M.L., Morsello, C., Oliveira, B., & Desbiez, A.L.J. (2021). Using a conflict framework analysis to help beekeepers and Giant Armadillos (*Priodontes maximus*) coexist. *Frontiers in Conservation Science, 2*, 696435. https//doi.org/10.3389/fcosc.2021.696435

Cattaneo, C., Maderna, E., Rendinelli, A., & Gibelli, D. (2015). Animal experimentation in forensic sciences: How far have we come? *Forensic Science International, 254*, e29–e35.

Catts, E.P., & Goff, M.L. (1992). Forensic entomology in criminal investigations. *Annual Review of Entomology, 37*, 253–272.

Cazzolla Gatti, R., Velichevskaya, A., Gottesman, B., & Davis, K. (2021). Grey wolf may show signs of self-awareness with the sniff test of self-recognition. *Ethology Ecology & Evolution, 33*(4), 444–467.

Ceballos-Lascurain, H. (1996). *Tourism, Ecotourism, and Protected Areas: The State of Nature-Based Tourism Around the World and Guidelines for its Development*. IUCN.

Chadwick, R., & Schroeder, D. (2002). *Applied Ethics: Critical Concepts in Philosophy*. Routledge.

Chang, C.C., Cheng, G.J.Y., Le Nghiem, T.P., Song, X.P., Oh, R.R.Y., Richards, D.R. et al. (2020). Social media, nature, and life satisfaction: Global evidence of the biophilia hypothesis. *Scientific Reports, 10*(1), 1–8.

Chavaly, D., & Naachimuthu, K.P. (2020). Human nature connection and mental health: What do we know so far? *Indian Journal of Health and Wellbeing, 11*(1–3), 84–92.

Chen, J., & Ten Cate, C. (2015). Zebra finches can use positional and transitional cues to distinguish vocal element strings. *Behavioural Processes, 117*, 29–34, https://dx.doi.org/10.1016/j.beproc.2014.09.004

Cheney, D.L., & Seyfarth, R.M. (1990). *How Monkeys See the World: Inside the Mind of Another Species*. University of Chicago Press.

Chesterton, R.N., Pfeiffer, D.U., Morris, R.S., & Tanner, C.M. (1989). Environmental and behavioural factors affecting the prevalence of foot lameness in New Zealand dairy herds – A case-control study. *New Zealand Veterinary Journal, 37*(4), 135–142.

Chiyo, P.I., Cochrane, E.P., Naughton, L., & Basuta, G.I. (2005). Temporal patterns of crop raiding by elephants: A response to changes in forage quality or crop availability? *African Journal of Ecology, 43*, 48–55.

Chiyo, P.I., Lee, P.C., Moss, C.J., Archie, E.A., Hollister-Smith, J.A., & Alberts, S.C. (2011). No risk, no gain: Effects of crop raiding and genetic diversity on body size in male elephants. *Behavioral Ecology, 22*, 552–558.

Chomsky, N. (1986). *Knowledge of Language: Its Nature, Origin, and Use*. Greenwood.

Chomsky, N. (2014). *The Minimalist Program*. MIT Press.

Chopik, W.J., & Weaver, J.R. (2019). Old dog, new tricks: Age differences in dog personality traits, associations with human personality traits, and links to important outcomes. *Journal of Research in Personality, 79*, 94–108. https://doi.org/10.1016/j.jrp.2019.01.005

CITES. (2021). *The CITES Species*. https://cites.org/eng/disc/species.php

Clark, G. (2021, 28 June). Draught wines: French vineyards rediscover the power of horses. *The Guardian*. www.theguardian.com/environment/2021/jun/28/draught-wines-french-vineyards-rediscover-the-power-of-horses

Claxton, A.M. (2011). The potential of the human–animal relationship as an environmental enrichment for the welfare of zoo-housed animals. *Applied Animal Behaviour Science, 133*(1–2), 1–10. https://doi.org/10.1016/j.applanim.2011.03.002

Clayton, D.F., Balakrishnan, C.N., & London, S.E. (2009). Integrating genomes, brain and behavior in the study of songbirds. *Current Biology, 19*(18), R865–R873.

Cloutier, S., Panksepp, J., & Newberry, R.C. (2012). Playful handling by caretakers reduces fear of humans in the laboratory rat. *Applied Animal Behaviour Science, 140*(3–4), 161–171. https://doi.org/10.1016/j.applanim.2012.06.001

Cohen, S., & Wills, T.A. (1985). Stress, social support, and the buffering hypothesis. *Psychological Bulletin, 98*, 310–357.

Cole, J., & Fraser, D. (2018). Zoo animal welfare: The human dimension. *Journal of Applied Animal Welfare Science, 21*(1), 49–58. https://doi.org/10.1080/10888705.2018.1513839

Coleman, G.J. (2004). Personnel management in agricultural systems. In G.J. Benson & E. Bernard (Eds.), *The Well-Being of Farm Animals: Challenges and Solutions* (pp. 167–181). Blackwell.

Coleman, G.J. (2018). Public animal welfare discussions and outlooks in Australia. *Animal Frontiers, 8*, 14–19.

Coleman, G.J., & Hemsworth, P.H. (2014). Training to improve stockperson beliefs and behaviour towards livestock enhances welfare and productivity. *Revue Scientifique et Technique, 33*(1), 131–137. https://doi.org/10.20506/RST.33.1.2257

Coleman, G.J., Hemsworth, P.H., & Hay, M. (1998). Predicting stockperson behaviour towards pigs from attitudinal and job-related variables and empathy. *Applied Animal Behaviour Science, 58*(1–2), 63–75.

Coleman, G.J., Hemsworth, P.H., Hay, M., & Cox, M. (2000). Modifying stockperson attitudes and behaviour towards pigs at a large commercial farm. *Applied Animal Behaviour Science, 66*, 11–20.

Coleman, G.J., Rice, M., & Hemsworth, P.H. (2012). Human-animal relationships at sheep and cattle abattoirs. *Animal Welfare, 21*, 15–21.

Collado, E.B., Martín, P.T., & Serena, O.C. (2023). Mapping human–animal interaction studies: A bibliometric analysis. *Anthrozoös, 36*(1), 137–157.

Coman, I.A., Cooper-Norris, C.E., Longing, S., & Perry, G. (2022). It is a wild world in the city: Urban wildlife conservation and communication in the age of COVID-19. *Diversity, 14*(7), 539.

Conde, D.A., Flesness, N., Colchero, F., Jones, O.R., & Scheuerlein, A. (2011). An emerging role of zoos to conserve biodiversity. *Science, 331*(6023), 1390–1391.

Constantine, R., Brunton, D.H., & Dennis, T. (2004). Dolphin-watching tour boats change bottlenose dolphin (*Tursiops truncatus*) behaviour. *Biological Conservation, 117*(3), 299–307.

Coppinger, R.P., & Coppinger, L. (2001). *Dogs: A New Understanding of Canine Origin, Behavior and Evolution*. University of Chicago Press.

Cormier, L. (2003). *Kinship with Monkeys: The Guajà Foragers of Eastern Amazonia.* Columbia University Press.

Cornish, A., Wilson, B., Raubenheimer, D., & McGreevy, P. (2018). Demographics regarding belief in non-human animal sentience and emotional empathy with animals: A pilot study among attendees of an animal welfare symposium. *Animals, 8*(10), 174.

Corr, C.A. (1999). Enhancing the concept of disenfranchised grief. *Omega, 38*(1), 1–20. https://doi.org/10.2190/LD26-42A6-1EAV-3MDN

Corrêa, G.F., Barcelos, A.M., & Mills, D.S. (2021). Dog-related activities and human well-being in Brazilian dog owners: A framework and cross-cultural comparison with a British study. *Science Progress, 104*(4), 00368504211050277.

Correia-Caeiro, C., Guo, K., & Mills, D. (2020). Perception of dynamic facial expressions of emotion between dogs and humans. *Animal Cognition, 23*, 465–476. https://doi.org/10.1007/s10071-020-01348-5

Correia-Caeiro, C., Guo, K., & Mills, D. (2021). Bodily emotional expressions are a primary source of information for dogs, but not for humans. *Animal Cognition, 24*, 267–279. https://doi.org/10.1007/s10071-021-01471-x

Costa, P.T., & McCrae, R.R. (1992). The five-factor model of personality and its relevance to personality disorders. *Journal of Personality Disorders, 6*, 343–359.

Courchamp, F., Jaric, I., Albert, C., Meinard, Y., Ripple, W.J., & Chapron, G. (2018). The paradoxical extinction of the most charismatic animals. *PLOS Biology, 16*(4), e2003997.

Cox, D.T.C., & Gaston, K.J. (2016). Urban bird feeding: Connecting people with nature. *PLOS ONE, 11*(7), e0158717. https://doi.org/10.1371/JOURNAL.PONE.0158717

Cransberg, P.H., Hemsworth, P.H., & Coleman, G.J. (2000). Human factors affecting the behaviour and productivity of commercial broiler chickens. *British Poultry Science, 41*, 272–279.

Crockford, C., Wittig, R.M., & Zuberbühler, K. (2017). Vocalizing in chimpanzees is influenced by social-cognitive processes. *Science Advances, 3*(11), e1701742.

Cronin, G.M., Rault, J.L., & Glatz, P.C. (2014). Lessons learned from past experience with intensive live-stock management systems. *Revue Scientifique et Technique, 33*, 139–151. https://doi.org/10.20506/rst.33.1.2256

Cuaya, L.V., Hernández-Pérez, R., & Concha, L. (2016). Our faces in the dog's brain: Functional imaging reveals temporal cortex activation during perception of human faces. *PLOS ONE, 11*(3), e0149431. https://doi.org/10.1371/journal.pone.0149431

Cummings, C.R., Lea, M.A., & Lyle J.M. (2019). Fur seals and fisheries in Tasmania: An integrated case study of human-wildlife conflict and coexistence. *Biological Conservation, 236*, 532–542.

Curb, L.A., Abramson, C.I., Grice, J.W., & Kennison, S.M. (2015). The relationship between personality match and pet satisfaction among dog owners. *Anthrozoös, 26*(3), 395–404. https://doi.org/10.2752/175303713X13697429463673

Curry, B.A., Donaldson, B., Vercoe, M., Filippo, M., & Zak, P.J. (2015). Oxytocin responses after dog and cat interactions depend on pet ownership and may affect interpersonal trust. *Human-Animal Interaction Bulletin, 3*(2), 56–71.

Curtin, S. (2009). Wildlife tourism: The intangible, psychological benefits of human-wildlife encounters. *Current Issues in Tourism, 12*(5–6), 451–474.

Curtin, S. (2010). Managing the wildlife tourism experience: The importance of tour leaders. *International Journal of Tourism Research, 12*(3), 219–236.

Curtin, S. (2013). Lessons from Scotland: British wildlife tourism demand, product development and destination management. *Journal of Destination Marketing & Management, 2*(3), 196–211.

Dabritz, H.A., & Conrad, P.A. (2010). Cats and toxoplasma: Implications for public health. *Zoonoses and Public Health, 57*(1), 34–52.

Dadds, M.R., Whiting, C., & Hawes, D.J. (2006). Associations among cruelty to animals, family conflict and psychopathic traits in childhood. *Journal of Interpersonal Violence, 21*(3),411–429. https://doi.org/10.1177/0886260505283341

Davis, E., Davies, B., Wolfe, R., Raadsveld, R., Heine, B., Thomason, P. et al. (2009). A randomized controlled trial of the impact of therapeutic horse riding on the quality of life, health, and function of children with cerebral palsy. *Developmental Medicine and Child Neurology, 51*, 111–119. https://doi.org/10.1111/j.1469-8749.2008.03245.x

Davis, H., Taylor, A.A., & Norris, C. (1997). Preference for familiar humans by rats. *Psychonomic Bulletin & Review, 4*(1), 118–120. https://doi.org/10.3758/BF03210783

De Risio, L., Bhatti, S., Muñana, K., Penderis, J., Stein, V., Tipold, A. et al. Diagnostic approach to epilepsy in dogs. *BMC Veterinary Research, 11*(1), 1–11.

de Waal, F.B. (2019). Fish, mirrors, and a gradualist perspective on self-awareness. *PLOS Biology, 17*(2), e3000112.

Debracque, C., Clay, Z., Grandjean, D., & Gruber, T. (2022). Humans recognize affective cues in primate vocalizations: Acoustic and phylogenetic perspectives. *bioRxiv*, 2022–01.

Décory, M.S.M. (2019). A universal definition of 'domestication' to unleash global animal welfare progress. *dA. Derecho Animal (Forum of Animal Law Studies)* 10(2), 39–55. https://doi.org/10.5565/rev/da.424

DeGue, S., & DiLillo, D. (2009). Is animal cruelty a 'red flag' for family violence? Investigating co-occurring violence toward children, partners, and pets. *Journal of Interpersonal Violence, 24*(6), 1036–1056. https://doi.org/10.1177/0886260508319362

Deldalle, S., & Gaunet, F. (2014). Effects of 2 training methods on stress-related behaviours of the dog (*Canis familiaris*) and on the dog–owner relationship. *Journal of Veterinary Behavior: Clinical Applications and Research, 9*(2), 58–65. https://doi.org/10.1016/j.jveb.2013.11.004

DeLoache, J.S., Pickard, M.B., & LoBue, V. (2011). How very young children think about animals. In P. McCradle, S. McCune, J.A. Griffin, & V. Maholmes (Eds.), *How Animals Affect Us: Examining the Influence of Human-Animal Interaction on Child Development and Human Health* (pp. 85–99). American Psychological Association

Demaree, H.A., Everhart, D.E., Youngstrom, E.A., & Harrison, D.W. (2005). Brain lateralization of emotional processing: Historical roots and a future incorporating 'dominance'. *Behavioral and Cognitive Neuroscience Reviews, 4*(1), 3–20.

DeMello, M. (2012). *Animals and Society: An Introduction to Human-Animal Studies.* Columbia University Press.

Dennett, D.C. (1983). Intentional systems in cognitive ethology: The 'Panglossian paradigm' defended. *Behavioral and Brain Sciences, 6*(3), 343–390.

Descartes, R. (1637). *Discours de la méthode pour bien conduire sa raison, et chercher la vérité dans les sciences.* Quatrième partie. Hachette.

Descovich, K., Li, X., Sinclair, M., Wang, Y., & Phillips, C.J.C. (2019). The effect of animal welfare training on the knowledge and attitudes of abattoir stakeholders in China. *Animals, 9*(11). https://doi.org/10.3390/ani9110989

Désiré, L., Boissy, A., & Veissier, I. (2002). Emotions in farm animals: A new approach to animal welfare in applied ethology. *Behavioural Processes, 60*, 81–85.

Devitt, C., Kelly, P., Blake, M., Hanlon, A., & More, S.J. (2015). An investigation into the human element of on-farm animal welfare incidents in Ireland. *Sociologia Ruralis, 55*(4), 400–416.

Dewes, M.F. (1978). Some aspects of lameness of dairy herds. *New Zealand Veterinary Journal, 26*, 147–159.

Dewey, C.W., Davies, E.S., Xie, H., & Wakshlag, J.J. (2019). Canine cognitive dysfunction: Pathophysiology, diagnosis, and treatment. *Veterinary Clinics: Small Animal Practice, 49*(3), 477–499.

Dewsbury, D.A. (1992). Comparative psychology and ethology: A reassessment. *American Psychologist, 47*(2), 208.

Dickman, A. (2010). Complexities of conflict: The importance of considering social factors for effectively resolving human-wildlife conflict. *Animal Conservation, 13*(5), 458–466.

Dietsch, A.M., Manfredo, M.J., & Teel, T.A., (2017). Wildlife value orientations as an approach to understanding the social context of human-wildlife conflict. In C.M. Hill, A.D. Webber, & N.E.C. Priston (Eds.), *Understanding Conflicts About Wildlife: A Biosocial Approach* (pp. 107–126). Berghahn Books.

Dilks, D.D., Cook, P., Weiller, S.K., Berns, H.P., Spivak, M., & Berns, G.S. (2015). Awake fMRI reveals a specialized region in dog temporal cortex for face processing. *PeerJ, 3*, e1115. https://doi.org/10.7717/peerj.1115

Diop, M., & Fadiga, M.L. (2018). *Evaluation de la Contribution Economique des Equides de trait au Senegal*. Brooke. www.thebrooke.org/sites/default/files/Images/2%20to%201%20ratio/Countries/Senegal/Contribution%20%C3%A9conomique%20des%20%C3%A9quid%C3%A9s%20de%20trait%20au%20S%C3%A9n%C3%A9gal_Final_.pdf

Donham, K.J., Thelin, A., Thelin, A., & Donham, K.J. (2016). Psychosocial conditions in agriculture. In K.J. Donham & A. Thelin (Eds.), *Agricultural Medicine* (pp. 351–377). Wiley Blackwell. https://doi-org.uml.idm.oclc.org/10.1002/9781118647356.ch10

The Donkey Sanctuary. (2019). Under the skin: Update on the global crisis for donkeys and the people who depend on them. www.thedonkeysanctuary.org.uk/sites/uk/files/2019-12/under-the-skin-report-english-revised-2019.pdf

Dowsett, E., Delfabbro, P., & Chur-Hansen, A. (2020). Adult separation anxiety disorder: The human-animal bond. *Journal of Affective Disorders, 270*, 90–96.

Doyle, C. (2017). Captive wildlife sanctuaries: Definition, ethical considerations and public perception. *Animal Studies Journal, 6*(2), 55–85.

Dreschel, N.A., & Granger, D.A. (2005). Physiological and behavioral reactivity to stress in thunderstorm-phobic dogs and their caregivers. *Applied Animal Behaviour Science, 95*(3–4), 153–168. https://doi.org/10.1016/J.APPLANIM.2005.04.009

Driessen, M.M. (2021). COVID-19 restrictions provide a brief respite from the wildlife roadkill toll. *Biological Conservation, 256*, 109012.

Druce, H.C., Mackey, R.L., & Slotow, R. (2011). How immunocontraception can contribute to elephant management in small, enclosed reserves: Munyawana population as a case study. *PLOS ONE, 6*(12), e27952.

Drudi, D.I.N.O. (2000). Are animals occupational hazards. *Compensation and Working Conditions, 5*(3), 15–22.

Duan, C., & Hill, C.E. (1996). The current state of empathy research. *Journal of Counselling Psychology, 43*, 261–274.

Dunbar, R. (2000). On the origin of the human mind. In P. Carruthers & A. Chamberlain (Eds.), *Evolution and the Human Mind: Modularity, Language and Meta-Cognition* (pp. 238–253). University of Wisconsin Press.

Dundes, A. (Ed.). (1994). *The Cockfight: A Casebook*. University of Wisconsin Press.

Dupré, G., & Heidenreich, D. (2016). Brachycephalic syndrome. *Veterinary Clinics: Small Animal Practice, 46*(4), 691–707.

Ebinghaus, A., Ivemeyer, S., & Knierim, U. (2018). Human and farm influences on dairy cows′ responsiveness towards humans – a cross-sectional study. *PLOS ONE, 13*(12), e0209817.

Ekman, P., & Friesen, W.V. (1971). Constants across cultures in the face and emotion. Journal of Personality and Social Psychology, 17(2), 124.

Eddy, R.G., & Scott, C.P. (1980). Some observations on the incidence of lameness in dairy cattle in Somerset. *Veterinary Record, 106*(7), 140–144.

Eddy, T.J., Gallup, G.G., & Povinelli, D.J. (1993). Attribution of cognitive states to animals: Anthropomorphism in comparative perspective. *Journal of Social Issues, 49*, 87–87.

Ekman, P., & Cordaro, D. (2011). What is meant by calling emotions basic. *Emotion Review, 3*(4), 364–370.

Eliason, S.L. (2003). Illegal hunting and angling: The neutralization of wildlife law violations. *Society & Animals, 11*(3), 225–43.

Ellenberg, U., Setiawan, A.N., Cree, A., Houston, D.M., & Seddon, P.J. (2007). Elevated hormonal stress response and reduced reproductive output in Yellow-eyed penguins exposed to unregulated tourism. *General and Comparative Endocrinology, 152*(1), 54–63.

Endo, K., Yamasaki, S., Ando, S., Kikusui, T., Mogi, K., Nagasawa, M. et al. (2020). Dog and cat ownership predicts adolescents' mental well-being: A population-based longitudinal study. *International Journal of Environmental Research and Public Health, 17*(3), 884.

Engel, G.L. (1977). The need for a new medical model: A challenge for biomedicine. *Science, 196*(4286), 129–136.

Engel, G.L. (1978). The biopsychosocial model and the education of health professionals. *Annals of the New York Academy of Sciences, 310*, 169–181. https://doi.org/10.1111/j.1749-6632.1978.tb22070.x

Engel, G.L. (1981). The clinical application of the biopsychosocial model. *The Journal of Medicine and Philosophy: A Forum for Bioethics and Philosophy of Medicine, 6*(2), 101–124. Oxford University Press.

Engesser, S., Ridley, A.R., & Townsend, S.W. (2016). Meaningful call combinations and compositional processing in the southern pied babbler. *Proceedings of the National Academy of Sciences, 113*(21), 5976–5981.

English, K., Jones, L., Patrick, D., & Pasini-Hill, D. (2003). Sexual offender containment. *Annals of the New York Academy of Sciences, 989*, 411–427. https://doi.org/10.1111/j.1749-6632.2003.tb07322.x

Erdman, P., Mueller, M., Schroeder, K., Jenkins, J., & Hart, L. (2023). Navigating pathways for careers in human-animal interaction: A guide for advisors and mentors. *Human–Animal Interactions*, 11, 1. https://doi.org/10.1079/hai.2023.0021

Evans, E.P. (1906). *The Criminal Prosecution and Capital Punishment of Animals*. W. Heinemann.

Evans, R., Lyons, M., Brewer, G., & Tucci, S. (2019). The purrfect match: The influence of personality on owner satisfaction with their domestic cat (*Felis silvestris catus*). *Personality and Individual Differences, 138*, 252–256. https://doi.org/10.1016/J.PAID.2018.10.011

Ewing, C.A., MacDonald, P.M., Taylor, M., & Bowers, M.J. (2007). Equine-facilitated learning for youths with severe emotional disorders: A quantitative and qualitative study. *Child & Youth Care Forum, 36*(1), 59–72. https://doi.org/10.1007/s10566-006-9031-x

Fa, J.E., Gusset, M., Flesness, N., & Conde, D.A. (2014). Zoos have yet to unveil their full conservation potential. *Animal Conservation, 17*(2), 97–100.

Fagre, A.C., Cohen, L.E., Eskew, E.A., Farrell, M., Glennon, E., Joseph, M.B. et al. (2022). Assessing the risk of human-to-wildlife pathogen transmission for conservation and public health. *Ecology Letters, 25*, 1534–1549. https://doi.org/10.1111/ele.14003

Faragó, T., Pongrácz, P., Miklósi, Á., Huber, L., Virányi, Z., & Range, F. (2010). Dogs' expectation about signalers' body size by virtue of their growls. *PLOS ONE, 5*(12). https://doi.org/ 10.1371/journal.pone.0015175

Faver, C.A., & Strand, E.B. (2003a). To leave or to stay? Battered women's concern for vulnerable pets. *Journal of Interpersonal Violence, 18*(12), 1367–1377. https://doi.org/10.1177/0886260503258028

Faver, C.A., & Strand, E.S. (2003b). Domestic violence and animal cruelty: Untangling the web of abuse. *Journal of Social Work Education, 39*(2), 237–253.

Febres, J., Brasfield, H., Shorey, R.C., Elmquist, J., Ninnemann, A., Schonbrun, Y.C., & Stuart, G.L. (2014). Adulthood animal abuse among men arrested for domestic violence. *Violence Against Women, 20*, 1059–1077. https://doi.org/10.1177/1077801214549641

Fennell, D.A. (2014). Exploring the boundaries of a new moral order for tourism's global code of ethics: An opinion piece on the position of animals in the tourism industry. *Journal of Sustainable Tourism, 22*(7), 983–996.

Fennell, D.A. (2020). *Ecotourism* (5th ed.). Routledge.

Fernandes, J.N., Hemsworth, P.H., Coleman, G.J. & Tilbrook, A.J. (2021). Costs and benefits of improving farm animal welfare. *Agriculture, 11*(2), 104.

Fine, A.H. (2019). *Handbook on Animal-AssistedTtherapy: Foundations and Guidelines for Animal-Assisted Interventions* (5th ed). Elsevier.

Fine, A.H., & Beck, A.M. (2015). Understanding our kinship with animals: Input for health care professionals interested in the human–animal bond. In A.H. Fine (Ed.), *Handbook on Animal-Assisted Therapy* (pp. 3–10). Elsevier.

Finka, L.R., Ward, J., Farnworth, M.J., & Mills, D.S. (2019). Owner personality and the well-being of their cats share parallels with the parent-child relationship. *PLOS ONE, 14*(2), 1–26. https://doi.org/10.1371/journal.pone.0211862

Finkemeier, M.A., Langbein, J., & Puppe, B. (2018). Personality research in mammalian farm animals: Concepts, measures, and relationship to welfare. *Frontiers in Veterinary Science, 5*(June), 1–15. https://doi.org/10.3389/fvets.2018.00131

Fiorillo, A., & Gorwood, P. (2020). The consequences of the COVID-19 pandemic on mental health and implications for clinical practice. *European Psychiatry, 63*(1), e32, 1–2. https://doi.org/10.1192/j.eurpsy.2020.35

Fisher, B., Polasky, S., & Sterner, T. (2011). Conservation and human welfare: Economic analysis of ecosystem services. *Environmental and Resource Economics, 48*(2), 151–159.

Fitch, W.T., Hauser, M.D., & Chomsky, N. (2005). The evolution of the language faculty: Clarifications and implications. *Cognition, 97*(2), 179–210.

Fitzgerald, A.J., Barrett, B.J., Gray, A., & Cheung, C.H. (2020). The connection between animal abuse, emotional abuse, and financial abuse in intimate relationships: Evidence from a nationally representative sample of the general public. *Journal of Interpersonal Violence, 37*(5–6), 2331–2353. https://doi.org/10.1177/0886260520939197

Food and Agriculture Organization of the United Nations (n.d.). *Country Participation by Round. World Programme for the Census of Agriculture*. www.fao.org/world-census-agriculture/wcarounds/results/en/

Ford, G., Guo, K., & Mills, D. (2019). Human facial expression affects a dog's response to conflicting directional gestural cues. *Behavioural Processes, 159*, 80–85. https://doi.org/ 10.1016/j.beproc.2018.12.022

Ford, J., Alleyne, E., Blake, E., & Somers, A. (2021). A descriptive model of the offence process for animal abusers: Evidence from a community sample. *Psychology, Crime & Law, 27*(4), 324–340. https://doi.org/10.1080/1068316X.2020.1798430

Forsyth, C.J., & Evans, R.D. (1998). Dogmen: The rationalisation of deviance. *Society & Animals, 6*(3), 203–218.

Franco, N.H., & Olsson, I.A.S. (2014). Scientists and the 3Rs: Attitudes to animal use in biomedical research and the effect of mandatory training in laboratory animal science. *Laboratory Animals, 48*(1), 50–60. https://doi.org/10.1177/0023677213498717

Franklin, A. (1999). *Animals and Modern Cultures: A Sociology of Human-Animal Relations in Modernity*. SAGE.

Fraser, D. (2005). *Animal Welfare and the Intensification of Animal Production. An Alternative Interpretation*. FAO readings in ethics. Food and Agriculture Organization of the United Nations, Rome. Available as a pdf on the FAO website: www.fao.org/ethics/ser_en.htm.

Fraser, D. (2008). *Understanding Animal Welfare: The Science in its Cultural Context*. Wiley-Blackwell.

Frey, R.G. (2011).Utilitarianism and animals. In T.L. Beauchamp and R.G. Frey (Eds.), *The Oxford Handbook of Animal Ethics* (pp. 172–197). Oxford University Press.

Friedmann, E., & Gee, N.R. (2017). Companion animals as moderators of stress responses: Implications for academic performance, testing and achievement. In N.R. Gee, A. Fine, & P. McCardle (Eds.), *How Animals Help Students Learn: Research and Practice for Educators and Mental Health Professionals* (pp. 98–110). Routledge, Taylor Francis Group.

Friedmann, E., Barker, S.B., & Allen, K.M. (2011). Physiological correlates of health benefits from pets. In P. McCardle, S. McCune, J.A. Griffin, & V. Maholmes (Eds.), *How Animals Affect Us: Examining the Influences of Human–Animal Interaction on Child Development and Human Health* (pp. 163–182). American Psychological Association. https://doi.org/10.1037/12301-009

Friedmann, E., Katcher, A.H., Lynch, J.J., & Thomas, S.A. (1980). Animal companions and one-year survival of patients after discharge from a coronary care unit. *Public Health Reports, 95*(4), 307–312.

Fritschi, L., Day, L., Shirangi, A., Robertson, I., Lucas, M., & Vizard, A. (2006). Injury in Australian veterinarians. *Occupational Medicine, 56*(3), 199–203. https://doi.org/10.1093/OCCMED/KQJ037

Fromm, E. (1973). *The Anatomy of Human Destructiveness*. Fawcett Crest.

Fugazza, C., Dror, S., Sommese, A., Temesi, A., & Miklósi, Á. (2021). Word learning dogs (*Canis familiaris*) provide an animal model for studying exceptional performance. *Scientific Reports, 11*(1), 1–9.

Funahashi, A., Gruebler, A., Aoki, T., Kadone, H., & Suzuki, K. (2014). Brief report: The smiles of a child with autism spectrum disorder during an animal-assisted activity may facilitate social positive behaviors – quantitative analysis with smile-detecting interface. *Journal of Autism and Developmental Disorders, 44*, 685–693. https://doi.org/10.1007/s10803-013-1898-4

Furton, K.G., & Myers, L.J. (2001). The scientific foundation and efficacy of the use of canines as chemical detectors for explosives. *Talanta, 54*, 487–500.

Gabriels, R.L., Pan, Z., Dechant, B., Agnew, J.A., Brim, N., & Mesibov, G. (2015). Randomized controlled trial of therapeutic horseback riding in children and adolescents with autism spectrum disorder. *Journal of the American Academy of Child and Adolescent Psychiatry, 54*(7), 541–549. https://doi.org/10.1016/j.jaac.2015.04.007

Gallup, G.G. (1970). Chimpanzees: Self-recognition. *Science, 167*(3914), 86–87.

Gallup, G.G. (1979). Self-awareness in primates: The sense of identity distinguishes man from most but perhaps not all other forms of life. *American Scientist, 67*(4), 417–421.

Gallup, G.G., & Anderson, J.R. (2018). The 'olfactory mirror' and other recent attempts to demonstrate self-recognition in non-primate species. *Behavioural Processes, 148*, 16–19.

Gallup, G.G., Anderson, J.R., & Shillito, D.J. (2002). The mirror test. In M. Bekoff, C. Allen, & G.M. Burghardt (Eds.), *The Cognitive Animal: Empirical and Theoretical Perspectives on Animal Cognition* (pp. 325–333). MIT Press.

Gallup, G.G., & Beckstead, J.W. (1988). Attitudes toward animal research. *American Psychologist, 43*(6), 474.

Gallup Jr, G.G., & Beckstead, J.W. (1989). Attitudes toward animal research. ILAR Journal, *31*(2), 13–15.

Gallup, G.G., McClure, M.K., Hill, S.D., & Bundy, R.A. (1971). Capacity for self-recognition in differentially reared chimpanzees. *Psychological Record, 21*(1), 69–74.

García-Gómez, A., Risco, M.L., Rubio, J.C., Guerrero, E., & García-Peña, I.M. (2013). Effects of a program of adapted therapeutic horse-riding in a group of autism spectrum disorder children. *Electronic Journal of Research in Educational Psychology, 12*(1), 107–128. https://doi.org/10.14204/ejrep.32.13115

Gardner, B.T., & Gardner, R.A. (1975). Evidence for sentence constituents in the early utterances of child and chimpanzee. *Journal of Experimental Psychology: General, 104*(3), 244.

Gardner, R.M. (2011). *Practical Crime Scene Processing and Investigation.* CRC Press.

Garnett, T., Appleby, M.C., Balmford, A., Bateman, I.J., Benton, T.G., Bloomer, P. et al. (2013). Sustainable intensification in agriculture: Premises and policies. *Science, 341*, 33–34.

Gazzaniga, M.S., Ivry, R.B., & Mangun, G.R. (2014). *Cognitive Neuroscience: The Biology of the Mind* (4th ed.). W.M. Norton.

Gee, N.R., Crist, E.N., & Carr, D.N. (2010). Preschool children require fewer instructional prompts to perform a memory task in the presence of a dog. *Anthrozoös, 23*(2), 173–184. https://doi.org/10.2752/175303710X12682332910051

Gee, N.R., Fine, A.H., Esposito, L., & McCune, S. (2017). Creating an atmosphere of acceptance for hai in education – future directions. In A. Fine & N. Gee. (eds.). *How Animals Help Students Learn: Research and Practice for Educators and Mental-Health Professionals* (pp. 12–26). Routledge.

Gee, N.R., Friedmann, E., Stendahl, M., Fisk, A. & Coglitore, V. (2014). Heartrate variability during a working memory task: Does touching a dog or person effect the response. *Anthrozoös, 27*(4), 513–528.

Gee, N.R., Gould, J.K., Swanson, C.C., & Wagner, A.K., (2012). Preschoolers categorize animate objects better in the presence of a dog. *Anthrozoös, 25*(2), 187–198.

Gee, N.R., Griffin, J.A., & McCardle, P. (2017). Human–animal interaction research in school settings: Current knowledge and future directions. *Aera Open, 3*(3), 2332858417724346.

Gee, N.R., Harris, S.L., & Johnson, K.L. (2007). The role of therapy dogs in speed and accuracy to complete motor skills tasks for preschool children. *Anthrozoös, 20*(2), 375–386. https://doi.org/10.2752/089279307X245509

Gee, N.R., & Mueller, M.K. (2019). A systematic review of research on pet ownership and animal interactions among older adults. *Anthrozoös, 32*(2), 183–207.

Gee, N.R., Rodriguez, K.E., Fine, A.H., & Trammell, J.P. (2021). Dogs supporting human health and well-Being: A biopsychosocial approach. *Frontiers in Veterinary Science, 8*, 630465. https://doi.org/10.3389/fvets.2021.630465

Gee, N.R., Sherlock, T.R., Bennett, E.A., & Harris, S.L. (2009). Pre-schoolers adherence to instructions as a function of presence of a dog and motor skills task. *Anthrozoös 22*(3), 267–276. https://doi.org/10.2752/175303709X457603

Geiger, M., Hockenhull, J., Buller, H., Tefera Engida, G., Getachew, M., Burden, F.A., & Whay, H.R. (2020). Understanding the attitudes of communities to the social, economic, and cultural importance of working donkeys in rural, peri-urban, and urban areas of Ethiopia. *Frontiers in Veterinary Science, 7*, 60.
Germonpré, M., Sablin, M.V., Stevens, R.E., Hedges, R.E., Hofreiter, M., Stiller, M. et al. (2009). Fossil dogs and wolves from Palaeolithic sites in Belgium, the Ukraine and Russia: Osteometry, ancient DNA and stable isotopes. *Journal of Archaeological Science, 36*(2), 473–490.
Gillon, R. (1998). Bioethics: Overview, in R. Chadwick (Ed.), *Encyclopedia of Applied Ethics* (pp. 305–317). Academic Press.
Glocker, M.L., Langleben, D.D., Ruparel, K., Loughead, J.W., Gur, R.C., & Sachser, N. (2009). Baby schema in infant faces induces cuteness perception and motivation for caretaking in adults. *Ethology, 115*(3), 257–263.
Gobbo, E., & Zupan, M. (2020). Dogs' sociability, owners' neuroticism and attachment style to pets as predictors of dog aggression. *Animals, 10*(2), 315. https://doi.org/10.3390/ani10020315
Goodin, R. (1993). Utility and the Good. In P. Singer (Ed.). *A Companion to Ethics* (pp. 241–248). Blackwell.
Gosling, S.D., & John, O.P. (1999). Personality dimensions in nonhuman animals: A cross-species review. *Current Directions in Psychological Science, 8*(3), 69–75. https://doi.org/10.1111/1467-8721.00017
Gourkow, N., & Fraser, D. (2006). The effect of housing and handling practices on the welfare, behaviour and selection of domestic cats (*Felis sylvestris catus*) by adopters in an animal shelter. Animal Welfare, 15(4), 371–377. http://206.223.197.179/docs/2011/ASSY/20111121_113/808_CCF11152011_00001.pdf
Graeub, B.E., Chappell, M.J., Wittman, H., Ledermann, S., Kerr, R.B., & Gemmill-Herren, B. (2016). The state of family farms in the world. *World Development, 87*, 1–15.
Graham, K.E., & Hobaiter, C. (2023). Towards a great ape dictionary: Inexperienced humans understand common nonhuman ape gestures. *PLOS Biology, 21*(1), e3001939.
Graham, L., & Gosling, S. (2009). Temperament and personality in working dogs. In W. Helton (Ed.), *Canine Ergonomics: The Science of Working Dogs* (pp. 63–81). Taylor & Francis.
Green, J.D., Coy, A.E., & Mathews, M.A. (2018). Attachment anxiety and avoidance influence pet choice and pet-directed behaviors. *Anthrozoös, 31*(4), 475–494.
Green, V.M., & Gabriel, K.I. (2020). Researchers' ethical concerns regarding habituating wild-nonhuman primates and perceived ethical duties to their subjects: Results of an online survey. *American Journal of Primatology, 82*(9). https://doi.org/10.1002/ajp.23178
Guan, D.W., Zhang, X.G., Zhao, R., Lu, B., Han, Y., Hou, Z.H., & Jia, J.T. (2007). Diverse morphological lesions and serious arrhythmias with hemodynamic insults occur in the early myocardial contusion due to blunt impact in dogs. *Forensic Science International, 166*(1), 49–57.
Guilherme Fernandes, J., Olsson, I.A.S., & Vieira de Castro, A.C. (2017). Do aversive-based training methods actually compromise dog welfare? A literature review. *Applied Animal Behaviour Science, 196*(July), 1–12. https://doi.org/10.1016/j.applanim.2017.07.001
Gulati, S., Karanth, K.K., Le, N.A., & Noack, F. (2021). Human casualties are the dominant cost of human–wildlife conflict in India. *Proceedings of the National Academy of Sciences, 118*(8), e1921338118.
Guo, K. (2012). Holistic gaze strategy to categorize facial expression of varying intensities. *PLOS ONE, 7*(8), e42585. https://doi.org/10.1371/journal.pone.0042585
Guo, K., Li, Z., Yan, Y., & Li, W. (2019). Viewing heterospecific facial expressions: An eye-tracking study of human and monkey viewers. *Experimental Brain Research, 237*, 2045–2059.
Guo, K., Meints, K., Hall, C., Hall, S., & Mills, D. (2009). Left gaze bias in humans, rhesus monkeys and domestic dogs. *Animal Cognition, 12*, 409–418. https://doi.org/10.1007/ s1007 1-008-0199-3
Guo, K., Smith, C., Powell, K., & Nicholls, K. (2012). Consistent left gaze bias in processing different facial cues. *Psychological Research, 76*, 263–269. https://doi.org/10.1007/ s00426-011-0340-9
Guo, K., Tunnicliffe, D., & Roebuck, H. (2010). Human spontaneous gaze patterns in viewing of faces of different species. *Perception, 39*, 533–542. https://doi.org/10.1068/p6517

Gupta, M. (2008). Functional links between intimate partner violence and animal abuse: Personality features and representations of aggression. *Society & Animals: Journal of Human–Animal Studies, 16*, 223–242. https://doi.org/10.1163/156853008X323385

Guthrie, S.E. (1997). Anthropomorphism: A definition and a theory. In R. Mitchell, N. Thompson and L. Miles (Eds.), *Anthropomorphism, Anecdotes, and Animals* (pp. 50–58). State University of New York Press.

Hall, S.S., Gee, N.R., & Mills, D.S. (2016). Children reading to dogs: A systematic review of the literature. *PLOS ONE, 11*(2), e:0149759. https://doi.org/10.1371/journal.pone.0149759

Hamm, M., & Mitchell, R.W. (1997). The interpretation of animal psychology: Anthropomorphism or behavior reading? *Behaviour, 134*(3–4), 173–204.

Hammerschmidt, K., & Fischer, J. (2019). Baboon vocal repertoires and the evolution of primate vocal diversity. *Journal of Human Evolution, 126*, 1–13.

Hansen, B.G. and Østerås, O. (2019). Farmer welfare and animal welfare – Exploring the relationship between farmer's occupational well-being and stress, farm expansion and animal welfare. *Preventative Veterinary Medicine, 170*, 104741.

Hare, B., Plyusnina, I., Ignacio, N., Schepina, O., Stepika, A., Wrangham, R. et al. (2005). *Current Biology, 15*, 226–230.

Hare, B., & Tomasello, M. (2005). Human-like social skills in dogs. *Trends in Cognitive Sciences, 9*, 439–444. https://doi.org/10.1016 /j.tics.2005.07.003

Hare, B., Wobber, V., & Wrangham, R. (2012).The self-domestication hypothesis: Evolution of bonobo psychology is due to selection against aggression. *Animal Behaviour, 83*, 573–585.

Harris, A., & Williams, J.M. (2017). The impact of a horse-riding intervention on the social functioning of children with autism spectrum disorder. *International Journal of Environmental Research and Public Health, 14*(7), 776. https://doi.org/10.3390/ijerph14070776

Harrison, M.A., & Hall, A.E. (2010). Anthropomorphism, empathy, and perceived communicative ability vary with phylogenetic relatedness to humans. *Journal of Social, Evolutionary, and Cultural Psychology, 4*(1), 34.

Haskell, N.H. (2007). Forensic entomology. *Medicolegal Investigation of* Death, *Guidelines for the* Application of *Pathology to Crime I*nvestigation. 4th ed. Thomas, Illinois, 149–173.

Hatfield, E., Cacioppo, J.T., & Rapson, R.L. (1992). Primitive emotional contagion. In M.S. Clark (Ed.), *Emotion and Social Behavior* (pp. 151–177). SAGE.

Hauser, M.D., Chomsky, N., & Fitch, W.T. (2002). The faculty of language: What is it, who has it, and how did it evolve? *Science, 298*(5598), 1569–1579.

Haverbeke, A., Laporte, B., Depiereux, E., Giffroy, J.M., & Diederich, C. (2008). Training methods of military dog handlers and their effects on the team's performances. *Applied Animal Behavioural Science, 113*(1–3), 110–112.

Havlíček, F. (2015). Waste management in hunter-gatherer communities. *Journal of Landscape Ecology, 8*(2), 47–59.

Hawkins, R., & Williams, J., & Scottish Society for the Prevention of Cruelty to Animals (SSPCA). (2017). Assessing Effectiveness of a Nonhuman Animal Welfare Education Program on Primary School Children. https://doi.org/10.1080/10888705.2017.1305272.

Hawkins, R.D., & Williams, J.M. (2016). Children's beliefs about animal minds (Child-BAM): Associations with positive and negative child–animal interactions. *Anthrozoös, 29*(3), 503–519.

Hawkins, R.D., & Williams, J.M. (2017). Childhood attachment to pets: Associations between pet attachment, attitudes to animals, compassion, and humane behaviour. *International Journal of Environmental Research and Public Health, 14*(5), 490.

Hawkins, R.D., & Williams, J.M. (2019). The development and pilot evaluation of a 'serious game' to promote positive child-animal interactions. *Human–Animal Interaction Bulletin, 8*(2).

Hawkins, R.D., Hawkins, E.L., & Tip, L. (2021). 'I can't give up when I have them to care for': People's experiences of pets and their mental health. *Anthrozoös,* 34(4), 543–562.

Hawkins, R.D., McDonald, S.E., O'Connor, K., Matijczak, A., Ascione, F.R., & Williams, J.H. (2019). Exposure to intimate partner violence and internalizing symptoms: The moderating effects of positive relationships with pets and animal cruelty exposure. *Child Abuse & Neglect, 98*, 104166.

Hawkins, R.D., Scottish S.P.C.A., & Williams, J.M. (2020). Children's attitudes towards animal cruelty: Exploration of predictors and socio-demographic variations. *Psychology, Crime & Law, 26*(3), 226–247.

Hawkins, R.D., Williams, J.M., & Scottish Society for the Prevention of Cruelty to Animals (SSPCA). (2017). Childhood attachment to pets: Associations between pet attachment, attitudes to animals, compassion, and humane behaviour. *International Journal of Environmental Research and Public Health, 14*(5), 490.

Hawley, F. (1993). The moral and conceptual universe of cockfighters: Symbolism and rationalization. *Society & Animals, 1*(2), 159–168.

Hayashi, T., Ishida, Y., Mizunuma, S., Kimura, A., & Kondo, T. (2009). Differential diagnosis between freshwater drowning and saltwater drowning based on intrapulmonary aquaporin-5 expression. *International Journal of Legal Medicine, 123*(1), 7–13.

Hayes, C. (1951). *The Ape in Our House*. Harper.

Hayes, J.E., McGreevy, P.D., Forbes, S.L., Laing, G., & Stuetz, R.M. (2018). Critical review of dog detection and the influences of physiology, training, and analytical methodologies. *Talanta, 185*, 499–512.

Heider, F., & Simmel, M. (1944). An experimental study of apparent behavior. *The American Journal of Psychology, 57*(2), 243–259.

Helton, W.S. (2009). Overview of scent detection work: Issues and opportunities. In W.S. Helton (Ed.), *Canine Ergonomics: The Science of Working Dogs* (pp. 83–97). CRC Press.

Hemsworth, P.H. (2003). Human–animal interactions in livestock production. *Applied Animal Behaviour Science, 81*, 185–198.

Hemsworth, P.H. (2019). Behavioural principles of pig handling. In T. Grandin (Ed.), *Livestock Handling and Transport* (5th ed., pp. 290–306). CAB International.

Hemsworth, P.H., Barnett, J.L., & Coleman, G.J. (1993). The human–animal relationship in agriculture and its consequences for the animal. *Animal Welfare, 2*(1), 33–51.

Hemsworth, P.H., Barnett, J.L., Coleman, G.J., & Hansen, C. (1989). A study of the relationships between the attitudinal and behavioural profiles of stockpersons and the level of fear of humans and reproductive performance of commercial pigs. *Applied Animal Behaviour Science, 23*(4), 301–314. www.sciencedirect.com/science/article/pii/0168159189900993

Hemsworth, P.H., Barnett, J.L., & Hansen, C. (1987). The influence of inconsistent handling by humans on the behaviour, growth and corticosteroids of young pigs. *Applied Animal Behaviour Science, 17*(3–4), 245–252. www.sciencedirect.com/science/article/pii/0168159187901493

Hemsworth, P.H., & Coleman, G.J. (2011). *Human–Livestock Interactions: The Stockperson and the Productivity and Welfare of Farmed Animals* (2nd ed.). CAB International.

Hemsworth, P.H., Coleman, G.J., Barnett, J.L., & Borg, S. (2000). Relationships between HAIs and productivity of commercial dairy cows. *Journal of Animal Science, 78*(11), 2821–2831. https://doi.org/10.2527/2000.78112821x

Hemsworth, P.H., Coleman, G.J., Barnett, J.L., Borg, S., & Dowling, S. (2002). The effects of cognitive behavioral intervention on the attitude and behaviour of stockpersons and the behaviour and productivity of commercial dairy cows. *Journal of Animal Science, 80*(1), 68–78. https://doi.org/10.2527/2002.80168x

Hemsworth, P.H., Brand, A., & Willems, P.J. (1981). The behavioural response of sows to the presence of human beings and their productivity. *Livestock Production Science, 8*, 67–74.

Hemsworth, P.H., Coleman, G.J., & Barnett, J.L. (1994a). Improving the attitude and behaviour of stockpersons towards pigs and the consequences on the behaviour and reproductive performance of commercial pigs. *Applied Animal Behaviour Science, 39*, 349–362.

Hemsworth, P.H., Coleman, G.J., Barnett, J.L., & Jones, R.B. (1994b). Fear of humans and the productivity of commercial broiler chickens. *Applied Animal Behaviour Science, 41*, 101–114.

Hemsworth, P.H., Coleman, G.J., Cox, M., & Barnett, J.L. (1994c). Stimulus generalisation: The inability of pigs to discriminate between humans on the basis of their previous handling experience. *Applied Animal Behaviour Science, 40*, 129–142.

Hemsworth, P.H., Coleman, G.C., Cransberg, P.H., & Barnett, J.L. (1996). *Human Factors and the Productivity and Welfare of Commercial Broiler Chickens*. Research Report on Chicken Meat Research and Development Council Project.

Hemsworth, P.H., Pedersen, V., Cox, M., Cronin, G.M., & Coleman, G.J. (1999). A note on the relationship between the behavioural response of lactating sows to humans and the survival of their piglets. *Applied Animal Behaviour Science, 65*, 43–52.

Hemsworth, P.H., Rice, M., Borg, S., Edwards, L.E., Ponnampalam, E.N., & Coleman, G.J. (2019). Relationships between handling, behaviour and stress in lambs at abattoirs. *Animal, 13*, 1287–1296.

Hemsworth, P.H., Sherwen, S.L., & Coleman, G.J. (2018). Human contact. In M. Appleby, I. Olsson and F. Galindo (Eds.), *Animal welfare* (3rd ed., pp. 294–314). CAB International.

Hensley, C., & Tallichet, S.E. (2005). Learning to be cruel? Exploring the onset and frequency of animal cruelty. *International Journal of Offender Therapy and Comparative Criminology, 49*(1), 37–47.

Hensley, C., Tallichet, S.E., & Dutkiewicz, E.L. (2010). Childhood bestiality: A potential precursor to adult interpersonal violence. *Journal of Interpersonal Violence, 25*(3), 557–567. https://doi.org/10.1177/0886260509360988

Herbeck, Y., Khantemirova, A., Antonov, V., Goncharova, N., Gulevich, R.G., Shepeleva, D., & Trut, L. (2017).Expression of the DNA methyltransferase genes in silver foxes experimentally selected for domestication. *Russian Journal of Genetics, 53*, 483–489.

Hergovich, A., Monshi., Semmler, G., & Zieglmayer, V. (2002). The effects of the presence of a dog in the classroom. *Anthrozoös, 15*(1), 37–50.

Hermann, N., Voß, C., & Menzel, S. (2013). Wildlife value orientations as predicting factors in support of reintroducing bison and of wolves migrating to Germany. *Journal for Nature Conservation, 21*, 125–132.

Herzog, H. (2011). *Some We Love, Some We Hate, Some We Wat: Why it's So Hard to Think Straight About Animals*. Harper Perennial.

Herzog, H. (2021, 24 May). Women dominate research on the human-animal bond. *Psychology Today.*

Herzog, H., Grayson, S., & McCord, D. (2015).Brief measures of the animal attitude scale. *Anthrozoös, 28*(1), 145–152, https://doi.org/10.2752/089279315X14129350721894

Herzog, H.A. (2014). Biology, culture and the origins pf pet-keeping. *Animal Behaviour and Cognition, 1*(3), 296–308. https://doi.org/10.12966/abc.08.06.2014.

Higginbottom, K. (2004). Managing impacts of wildlife tourism on wildlife. In K. Higginbottom (Ed.), *Wildlife Tourism: Impacts, Management and Planning* (pp. 211–229). Common Ground.

Hill, C.M. (2004). Farmers' perspectives of conflict at the wildlife-agriculture boundary: An African case study. *Human Dimensions of Wildlife, 9*(4), 279–286.

Hill, C.M. (2018). Crop foraging, crop losses and crop raiding. *Annual Review of Anthropology, 47*, 377–394. https//doi.org/10.1146/annurev-anthro-102317-050022

Hill, C.M. (2021). Conflict is integral to human-wildlife coexistence. *Frontiers in Conservation Science, 2*, 734314. https://doi.org/10.3389/fcosc.2021.734314

Hill, E.M., LaLonde, C.M., & Reese, L.A. (2020). Compassion fatigue in animal care workers. *Traumatology, 26*(1), 96–108. https://doi.org/10.1037/trm0000218

Hill, S.D., Bundy, R.A., Gallup, G.G. Jr., & McClure, M.K. (1970). Responsiveness of young nursery reared chimpanzees to mirrors. *Proceedings of the Louisiana Academy of Sciences, 33*, 77–82.

Hirschenhauser, K., Meichel, Y., Schmalzer, S., & Beetz, A.M. (2017). Children love their pets: Do relationships between children and pets co-vary with taxonomic order, gender, and age? *Anthrozoös, 30*(3), 441–456.

Hitchens, P.L., Booth, R.H., Stevens, K., Murphy, A., Jones, B., & Hemsworth, L.M. (2021). The welfare of animals in Australian filmed media. *Animals, 11*(7), 1986.

Hobaiter, C., & Byrne, R.W. (2014). The meanings of chimpanzee gestures. *Current Biology, 24*(14), 1596–1600.

Hockett, C.F. (1960). The origin of speech. *Scientific American, 203*(3), 88–97.

Hodges, J. (2000). Why livestock, ethics and quality of life? In J. Hodges and I.K. Han (Eds.), *Livestock, Ethics and Quality of Life* (pp. 1–26). CABI.

Hoehl, S., Hellmer, K., Johansson, M., & Gredebäck, G. (2017). Itsy bitsy spider…: Infants react with increased arousal to spiders and snakes. *Frontiers in Psychology, 8*, 1710.

Höglin, A., Poucke, E. Van, Katajamaa, R., Jensen, P., Theodorsson, E., & Roth, L.S.V. (2021). Long-term stress in dogs is related to the human–dog relationship and personality traits. *Scientific Reports, 11*(1), 1–9. https://doi.org/10.1038/s41598-021-88201-y

Holoyda, B., & Newman, W. (2014). Zoophilia and the law: Legal responses to a rare paraphilia. *Journal of the American Academy of Psychiatry and the Law Online, 42*(4), 412–420.

Holoyda, B., Sorrentino, R., Friedman, S.H., & Allgire, J. (2018). Bestiality: An introduction for legal and mental health professionals. *Behavioral Sciences & the Law, 36*(6), 687–697.

Home, C., Bhatnagar, Y., & Vanak, A. (2018). A canine conundrum: Domestic dogs as invasive species and their impacts on wildlife in India. *Animal Conservation, 21*, 275–282.

Horowitz, A. (2009). Disambiguating the 'guilty look': Salient prompts to a familiar dog behaviour. *Behavioural Processes, 81*(3), 447–452.

Hosey, G. (2008). A preliminary model of human–animal relationships in the zoo. *Applied Animal Behaviour Science, 109*(2–4), 105–127. https://doi.org/10.1016/j.applanim.2007.04.013

Hosey, G., & Melfi, V. (2012). Human–animal bonds between zoo professionals and the animals in their care. *Zoo Biology, 31*(1), 13–26. https://doi.org/10.1002/ZOO.20359

Hosey, G., & Melfi, V. (2014). Human-animal interactions, relationships and bonds: A review and analysis of the literature. *International Journal of Comparative Psychology, 27*(1).

Hosey, G., & Melfi, V. (2019). *Anthrozoology: HAIs in Domesticated and Wild Animals*. Oxford University Press

Hosey, G.R. (2000). Zoo animals and their human audiences: What is the visitor effect? *Animal Welfare, 9*(4), 343–357.

Hostiou, N., Fagon, J., Chauvat, S., Turlot, A., Kling, F., Boivin, X., & Allain, C. (2017). Impact of precision livestock farming on work and human-animal interactions on dairy farms. *A Review. Bioscience, Biotechnology and Biochemistry,* 21, 1–8.

Hounsome, B., Edwards, R.T., Hounsome, N., & Edward-Jones, G. (2012). Psychological morbidity of farmers and non-farming population: Results from a UK survey. *Community Mental Health Journal, 48*, 503–510. https://doi.org/10.1007/s10597-011-9415-8

House of Commons Library. (2018). Badger culling in England. https://commonslibrary.parliament.uk/research-briefings/sn06837

Huang, Q.Y., Chen, Y.C., & Liu, S.P. (2012). Connexin 43, angiotensin II, endothelin 1, and type III collagen alterations in heart of rats having undergone fatal electrocution. *American Journal of Forensic Medicine and Pathology, 33*(3), 215–221.

Huber, A., Barber, A.L.A., Faragó, T., Müller, C.A., & Huber, L. (2017). Investigating emotional contagion in dogs (*Canis familiaris*) to emotional sounds of humans and conspecifics. *Animal Cognition, 20*, 703–715. https://doi.org/10.1007/s10071-017-1092-8

Huber, L., Racca, A., Scaf, B., Virányi, Z., & Range, F. (2013). Discrimination of familiar human faces in dogs (*Canis familiaris*). *Learning and Motivation, 44*, 258–269. https://doi.org/ 10.1016/j.lmot.2013.04.005

Humle, T., & Hill, C.M. (2016). People-primate interactions: Implications for primate conservation. In S. Wich & A. Marshall (Eds.), *An Introduction to Primate Conservation* (pp. 219–240). Oxford University Press.

Hunter, L.T.B., White, P., Henschel, P., Frank, L., Burton, C., Loveridge, A. et al. (2013). Walking with lions: Why there is no role for captive-origin lions Panthera leo in species restoration. *Oryx, 47*(1), 19–24.

Hunter, R.G., Gagnidze, K., McEwen, B.S., & Pfaff, D.W. (2015).Stress and the dynamic genome: Steroids, epigenetics, and the transposome. *Proceedings of the National Academy of Sciences of the United States of* America, *112*(22), 6828–6833.

Hursthouse, R. (2011).Virtue ethics and the treatment of animals. In T.L. Beauchamp & R.G. Frey (Eds.), *The Oxford Handbook of Animal Ethics* (pp. 119–143). Oxford University Press.

Hurt, A., & Smith, D.A. (2009). *Conservation Dogs. Canine ergonomics: The Science of Working Dogs*. CRC Press.

Hyun, G.J., Jung, T.W., Park, J.H., Kang, K.D., Kim, S.M., Son, Y.D. et al. (2016). Changes in gait balance and brain connectivity in response to equine-assisted activity and training in children with attention deficit hyperactivity disorder. *Journal of Alternative and Complementary Medicine, 22*(4), 286–293. https://doi.org/10.1089/acm.2015.0299

Intarapanich, N.P., McCobb, E.C., Reisman, R.W., Rozanski, E.A., & Intarapanich, P.P. (2016). Characterization and comparison of injuries caused by accidental and non-accidental blunt force trauma in dogs and cats. *Journal of Forensic Sciences, 61*(4), 993–999. https://doi.org/10.1111/1556-4029.13074

International Association for the Study of Pain (IASP). (2022, 3 January). Definition of pain. www.iasp-pain.org/resources/terminology/#pain

International Association of Human-Animal Interaction Organisations (IAHAIO). (2014–2018, updated 2020). The IAHAIO definitions of animal-assisted intervention and guidelines for wellness of animals involved in AAI. https://iahaio.org/wp/wp-content/uploads/2021/01/iahaio-white-paper-2018-english.pdf

The International Ecotourism Society. (2015). What is ecotourism? The definition. https://ecotourism.org/what-is-ecotourism/

International Labour Organization (ILO). (n.d.). What is forced labour, modern slavery and human trafficking. www.ilo.org/global/topics/forced-labour/definition/lang--en/index.htm

International Labour Organization (ILO), Brooke, & The Donkey Sanctuary. (2017). *Brick by Brick*. www.thebrooke.org/sites/default/files/Brooke%20News/Brick-by-Brick-Report.pdf

Ireland, J.L., Birch, P., Lewis, M., Mian, U., & Ireland, C.A. (2021). Animal abuse proclivity among women: Exploring callousness, sadism, and psychopathy traits. *Anthrozoös,* 35(1), 37–53. https://doi.org/10.1080/08927936.2021.1944560

Islam, A., & Towell, T. (2013). Cat and dog companionship and well-being: A systematic review. *International Journal of Applied Psychology, 3*(6), 149–155.

Jackendoff, R. (1993). *Patterns in the Mind: Language and Human Nature*. Harvester Wheatsheaf.

Jackendoff, R. (2002). *Foundations of Language: Brain, Meaning, Grammar, Evolution*. Oxford University Press.

Jackendoff, R. (2010). *Meaning and the Lexicon: The Parallel Architecture 1975–2010*. Oxford University Press.

Jadhav, S., & Barua, M. (2012). The elephant vanishes: Impact of human-elephant conflict on human well-being. *Health and Place, 18*, 1356–1365.

Janssen, L. Giemsch, L., Schmitz, R., Street, M., Van Dongen, S., Crobe, P.A. (2018). A new look at an old dog: Bonn-Oberkassel reconsidered. *Journal of Archaeological Science*, 92, 126–138. https://doi.org/10.1016\j.jas.2018.01.004

Jenkins, S.R., & DiGennaro Reed, F.D. (2013). An experimental analysis of the effects of therapeutic horseback riding on the behavior of children with autism. *Research in Autism Spectrum Disorders, 7*, 721–740. https://doi.org/10.1016/j.rasd.2013.02.008

Jensen, P. (2015).Adding 'epi-' to behaviour genetics: Implications for animal domestication. *Journal of Experimental Biology, 218*(1), 32–40.

Jeyaretnam, J., Jones, H., & Phillips, M. (2000). Disease and injury among veterinarians. *Australian Veterinary Journal, 78*(9), 625–629.

Jezierski, T., Adamkiewicz, E., Walczak, M., Sobczynska, M., Gorecka-Bruzda, A., Ensminger, J. et al. (2014). Efficacy of drug detection by fully-trained police dogs varies by breed, training level, type of drug and search environment. *Forensic Science International, 237*, 112–118.

John, O.P., & Srivastava, S. (1999). The Big-Five Trait Taxonomy: History, measurement, and theoretical perspectives. In L. Pervin & O.P. John (Eds.), *Handbook of Personality: Theory and Research* (2nd ed., pp. 102–138). Guilford Press.

Johnson, E.A., Portillo, A., Bennett, N.E., & Gray, P.B. (2021). Exploring women's oxytocin responses to interactions with their pet cats. *PeerJ, 9*, e12393.

Johnson, R.A., Albright, D.L., Marzolf, J.R., Bibbo, J.L., Yaglom, H.D., Crowder, S.M. et al. (2018). Effects of therapeutic horseback riding on post-traumatic stress disorder in military veterans.

*Military Medical Res*earch, *5*(1), 3. https://doi.org/10.1186/s40779-018-0149-6. PMID: 29502529; PMCID: PMC5774121.

Jones, A.C. (2008). *Development and Validation of a Dog Personality Questionnaire*. [Unpublished doctoral thesis], University of Texas at Austin.

Jones, D., Brace, C.L., Jankowiak, W., Laland, K.N., Musselman, L.E., Langlois, J.H. et al. (1995). Sexual selection, physical attractiveness, and facial neoteny: Cross-cultural evidence and implications [and comments and reply]. *Current Anthropology, 36*(5), 723–748.

Jones-Bitton, A., Best, C., MacTavish, J., Fleming, S., & Hoy, S. (2020). Stress, anxiety, depression, and resilience in Canadian farmers. *Social Psychiatry and Psychiatric Epidemiology, 55*, 229–236. https://doi.org/10.1007/s00127-019-01738-2

Jones-Engel, L., Engel, G.A., Schillaci, M.A., Rompis, A., Putra, A., Suaryana, K. et al. (2005). Primate to human retroviral transmission in Asia. *Emerging Infectious Diseases, 7*, 1028–1035.

Joye, Y., & De Block, A. (2011). 'Nature and I are two': A critical examination of the biophilia hypothesis. *Environmental Values, 20*(2), 189–215.

Judge, T.A., Rodell, J.B., Klinger, R.L., Simon, L.S., & Crawford, E.R. (2013). Hierarchical representations of the five-factor model of personality in predicting job performance: Integrating three organizing frameworks with two theoretical perspectives. *Journal of Applied Psychology, 98*(6), 875–925.

Kaiser, M., & Forsberg, E.M. (2001). Assessing fisheries – using an ethical matrix in a participatory process. *Journal of Agricultural and Environmental Ethics, 14*(2), 191–200.

Kaminski, J., Call, J., & Fischer, J. (2004). Word learning in a domestic dog: evidence for "fast mapping". *Science, 304*(5677), 1682–1683.

Kanamori, M., & Kondo, N. (2020). Suicide and types of agriculture: A time-series analysis in Japan. *Suicide and Life-Threatening Behavior, 50*(1), 122–137.

Kansky, R., Kidd, M., & Knight, A.T. (2014). Meta-analysis of attitudes toward damage-causing mammalian wildlife. *Conservation Biology, 28*(4), 924–938. https://doi.org/10.1111/cobi.12275

Kanthaswamy, S. (2015). Domestic animal forensic genetics – biological evidence, genetic markers, analytical approaches and challenges. *Animal Genetics, 46*(5), 473–484.

Kaplan, R., & Kaplan, S. (1989). *The Experience of Nature: A Psychological Perspective*. Cambridge University Press.

Kapustka, J., & Budzyńska, M. (2020). Does animal-assisted intervention work? Research review on the effectiveness of AAI with the use of different animal species. Human and Veterinary Medicine, 12(3), 135–141.

Kaulfuß, P., & Mills, D.S. (2008). Neophilia in domestic dogs (Canis familiaris) and its implication for studies of dog cognition. Animal Cognition, 11, 553–556.

Kauppinen, T., Valros, A. & Vesala, K.M. (2013). Attitudes of dairy farmers toward cow welfare in relation to housing, management and productivity. *Anthrozoös, 26*(3), 405–420.

Kauschke, C., & Hofmeister, C. (2002). Early lexical development in German: A study on vocabulary growth and vocabulary composition during the second and third year of life. *Journal of Child Language, 29*(4), 735–757.

Kavanagh, P.S., Signal, T.D., & Taylor, N. (2013). The Dark Triad and animal cruelty: Dark personalities, dark attitudes, and dark behaviors. *Personality and Individual Differences, 55*(6), 666–670. https://doi.org/10.1016/j.paid.2013.05.019

Kellert, S.R., & Wilson, E.O. (Eds.). (1993). *The Biophilia Hypothesis*. Island Press

Kenji, N., Ihama, Y., Fukasawa, M., Nagai, T., Fuke, C., & Miyazaki, T. (2012). Morphologic investigation of injury caused by locally applied negative pressure in a rat model. *Legal Medicine, 14*(1), 21–26.

Kenya National Bureau of Statistics Census Data. (2019). *Population and Housing Census*. (Vol. 4: Distribution of population by socio-economic characteristics). Ministry of Planning and National Development.

Kern, J.K., Fletcher, C.L., Garver, C.R., Mehta, J.A., Grannemann, B.D., & Knox, K.R. (2011). Prospective trial of equine-assisted activities in autism spectrum disorder. *Alternative Therapies in Health and Medicine, 17*(3), 14–20.

Kersken, V., Gómez, J.C., Liszkowski, U., Soldati, A., & Hobaiter, C. (2019). A gestural repertoire of 1- to 2-year-old human children: In search of the ape gestures. *Animal Cognition, 22*(4), 577–595.

Kertes, D.A., Liu, J., Hall, N.J., Hadad, N.A., Wynne, C.D.L., & Bhatt, S.S. (2017). Effect of pet dogs on children's perceived stress and cortisol stress response. *Social Development, 26*(2), 382–401. https://doi.org/10.111/sode.12203.

Khumalo, K.E., & Yung, L.A. (2013). Women, human-wildlife conflict and CBNRM: Hidden impacts and vulnerabilities in Kwandu Conservancy, Namibia. *Conservation & Society, 13*, 232–243.

Kiesler, S., Lee, S.L., & Kramer, A.D. (2006). Relationship effects in psychological explanations of non-human behavior. *Anthrozoös, 19*(4), 335–352.

Kinderman, P., Dunbar, R., & Bentall, R.P. (1998). Theory-of-mind deficits and causal attributions. *British Journal of Psychology, 89*(2), 191–204.

King, M.T.M., Matson, R.D., & DeVries, T.J. (2021). Connecting farmer mental health with cow health and welfare on dairy farms using robotic milking systems. *Animal Welfare, 30*, 25–38.

Kirnan, J., Siminerio, S., & Wong, Z. (2016). The impact of a therapy dog program on skills and attitudes towards reading. *Journal of Early Childhood Education, 44*, 637–651.

Kis, A., Bence, M., Lakatos, G., Pergel, E., Turcsán, B., Pluijmakers, J. et al. (2014). Oxytocin receptor gene polymorphisms are associated with human directed social behavior in dogs (*Canis familiaris*). *PLOS ONE, 9*, e83993.

Knight, B. (1992). Forensic science and animal rights. *Forensic Science International, 57*, 1–3.

Knight, J. (2010). The ready-to-view wild monkey. The convenience principle in Japanese wildlife tourism. *Annals of Tourism Research, 37*(3), 744–762.

Knight, S., Vrij, A., Cherryman, J., & Nunkoosing, K. (2004). Attitudes towards animal use and belief in animal mind. *Anthrozoös, 17*(1), 43–62.

Kohn, E. (2013). *How Forests Think: Toward an Anthropology Beyond the Human*. University of California Press.

Kollman J. (1905). *Neue Gedanken über das alter Problem von der Abstammung des Menschen. Correspondenz-Blatt der Dt. Ges. für Anthropologie*. (Leipzig), n.pub.

Kolstrup, C.L., & Hultgren, J. (2011). Perceived physical and psychosocial exposure and health symptoms of dairy farm staff and possible associations with dairy cow health. *Journal of Agricultural Safety and Health, 17*(2), 111–125.

Komar, D. (1999). The use of cadaver dogs in locating scattered, scavenged human remains: Preliminary field test results. *Journal of Forensic Science, 44*(2), 405–408.

Korsgaard, C.M. (2011). Interacting with animals: A Kantian account. In T.L. Beauchamp and R.G. Frey (Ed.), *The Oxford Handbook of Animal Ethics* (pp. 91–118). Oxford University Press.

Koster, J. (2021). Most dogs are not NATIVE dogs. *Integrative and Comparative Biology, 61*(1), 110–116. https://doi.org/10.1093/ICB/ICAB016

Köszegi, H., Fugazza, C., Magyari, L., Iotchev, I.B., Miklósi, Á., & Andics, A. (2023). Investigating responses to object-labels in the domestic dog (*Canis familiaris*). *Scientific Reports, 13*(1), 3150.

Kotrschal, K., & Ortbauer, B. (2003). Behavioural effects if the presence of a dog in the classroom. *Anthrozoös, 16*(2), 147–159.

Kruger, D.J. (2015). Non-mammalian infants requiring parental care elicit greater human caregiving reactions than superprecocial infants do. *Ethology, 121*(8), 769–774.

Kruger National Park. (n.d.). Elephant population numbers in Kruger. www.krugerpark.co.za/krugerpark-times-23-elephant-numbers-18006.html

Kubinyi, E., Virányi, Z., & Miklosi, A. (2007). Comparative social cognition: From wolf and dog to humans. *Comparative Cognition and Behavior Reviews, 2*, 26–46.

Kurdek, L.A. (2009). Pet dogs as attachment figures for adult owners. *Journal of Family Psychology, 23*(4), 439.

Kwong, M.J., & Bartholomew, K. (2011). 'Not just a dog': An attachment perspective on relationships with assistance dogs. *Attachment & Human Development, 13*(5), 421–436.

LaFollette, M.R., Riley, M.C., Cloutier, S., Brady, C.M., O'Haire, E., & Gaskill, B.N. (2020). Laboratory animal welfare meets human welfare: A cross-sectional study of professional quality of life, including compassion fatigue in laboratory animal personnel. *Frontiers in Veterinary Science, 7*, 114. https://doi.org/10.3389/fvets.2020.00114

Lagoni, L. (2011). Family-present euthanasia: Protocols for planning and preparing clients for the death of a pet. In C. Blazina, G. Boyraz, & D. Shen-Miller (Eds.), *The Psychology of the Human-Animal Bond* (pp. 181–202). Springer. https://doi.org/10.1007/978-1-4419-9761-6_11

Lambert, H., & Carder, G. (2019). Positive and negative emotions in dairy cows: Can ear postures be used as a measure?. *Behavioural Processes, 158*, 172–180.

Landau, B., Smith, L.B., & Jones, S.S. (1988). The importance of shape in early lexical learning. *Cognitive Development, 3*(3), 299–321.

Lange, A., Waiblinger, S., Heinke, A. Barth, K., Futschik, A., & Lurzel, S. (2020). Gentle interactions with restrained and free-moving cows: Effects on the improvement of the animal-human relationship. *PLOS ONE, 15*(11), e0242873.

Lanning, B.A., Baier, M.E., Ivey-Hatz, J., Krenek, N., & Tubbs, J.D. (2014). Effects of equine assisted activities on autism spectrum disorder. *Journal of Autism and Developmental Disorders, 44*(8), 1897–1907.

Larrère, C., & Larrère, R. (2000). Animal rearing as a contract? *Journal of Agricultural and Environmental Ethics, 12*, 65–78.

Lavorgna, B.F., & Hutton, V.E. (2019).Grief severity: A comparison between human and companion animal death. *Death Studies, 43*(8), 521–526. https://doi.org/10.1080/07481187.2018.1491485

Lawrence, E.A. (1994). Conflicting ideologies: Views of animal rights advocates and their opponents. *Society & Animals, 2*(2). https://doi.org/10.1163/156853094X00199

Le Roux, M.C., Swartz, L., & Swartz, E. (2014). The effect of an animal-assisted reading program on the reading rate, accuracy and comprehension of grade 3 students: A randomised control study. *Child & Youth Care, 43*, 655–673, https://doi.org/10.1007/s10566-014-9262-1.

Leconstant, C., & Spitz, E. (2022). Integrative model of human-animal interactions: A One Health–One Welfare systemic approach to studying HAI. *Frontiers in Veterinary Science, 9*, 656833–656833.

LeDoux, J.E. (2012). Evolution of human emotion: A view through fear. *Progress in Brain Research, 195*, 431–442.

Lee, P.C., Poole, J.H., Njiraini, N., Sayialel, C.N., & Moss, C.J. (2011). Male social dynamics: Independence and beyond. In C.J. Moss, H. Croze, & P.C. Lee (Eds.), *The Amboseli Elephants: A Long-Term Perspective on a Long-Lived Mammal* (pp. 260–271). University of Chicago Press.

Lehman, B.J., David, D.M., & Gruber, J.A. (2017). Rethinking the biopsychosocial model of health: Understanding health as a dynamic system. *Social and Personality Psychology Compass, 11*, e12328. https://doi.org/10.1111/spc3.12328

Leighty, K.A., Valuska, A.J., Grand, A.P., Bettinger, T.L., Mellen, J.D., Ross, S.R. et al. (2015). Impact of visual context on public perceptions of non-human primate performers. *PLOS ONE, 10*(2), e0118487.

Leon, A.F., Sanchez, J.A., & Romero, M.H. (2020). Association between attitude and empathy with the quality of human-livestock interactions. *Animals, 10*(8), 1304. https://doi.org/10.3390/ani10081304

Leong, K.M. (2009). The tragedy of becoming common: Landscape change and perceptions of wildlife. *Society and Natural Resources, 23*, 111–127.

Lesthaeghe, R. (2014). The second demographic transition: A concise overview of its development. *PNAS, 111*(51), 18112–18115. https://doi.org/10.1073/pnas.1420441111

Levine, D.W. (2005). Do dogs resemble their owners? A reanalysis of Roy and Christenfeld (2004). *Psychological Science, 16*(1), 83–84. https://doi.org/10.1111/J.0956-7976.2005.00785.X

Levinson, B. (1969). *Pet-Oriented Psychotherapy*. Charles C. Thomas.

Li, S., Sun, Y., Shan, D., Feng, B., Xing, J., Duan, Y. et al. (2013). Temporal profiles of axonal injury following impact acceleration traumatic brain injury in rats – a comparative study with diffusion tensor imaging and morphological analysis. *International Journal of Legal Medicine, 127*(1), 159–167.

Lim, M.M., & Young, L.J. (2006). Neuropeptidergic regulation of affiliative behavior and social bonding in animals. *Hormones and Behavior, 50*, 506–517.

Limond, J.A., Bradshaw, J.W.S., & Cormack, M.K.F. (1997). Behavior of children with learning disabilities interacting with a therapy dog. *Anthrozoös, 10*(2–3), 84–89. https://doi.org/10.2752/089279397787001139

Lindenmayer, J.M., & Kaufman, G.E. (2021). One Health and One Welfare. In T. Stephens (Ed.), *One Welfare in Practice* (pp. 1–30). CRC Press.

Linder, D.E., Santiago, S., & Halbreich, E.D. (2021). Is there a correlation between dog obesity and human obesity? Preliminary findings of overweight status among dog owners and their dogs. *Frontiers in Veterinary Science, 1*, 654617. https://doi.org/10.3389/fvets.2021.654617

Lit, L., Schweitzer, J.B., & Oberbauer, A.M. (2011). Handler beliefs affect scent detection dog outcomes. *Animal Cognition, 14*(3), 387–394.

Lizarralde, M. (2002). Ethnoecology of monkeys among the Bari of Venezuela: Perception, use and conservation. In A. Fuentes & L.D. Wolfe (Eds.), *Primates Face to Face: The Conservation Implications of Human-Nonhuman Primate Interconnections* (pp. 85–100). Cambridge University Press.

Lockwood, R. (2018). Animal hoarding: The challenge for mental health, law enforcement, and animal welfare professionals. *Behavioral Sciences and the Law, 36*(6), 698–716. https://doi.org/10.1002/bsl.2373

Lorenz, K. (1967). *So kam der Mensch auf den Hund* [Man meets dog]. dtv Verlagsgesellschaft.

Love, H.A. (2021). Best friends come in all breeds: The role of pets in suicidality. *Anthrozoös, 34*(2), 175–186. https://doi.org/10.1080/08927936.2021.1885144

Lynch, M.J., Long, M.A., & Stretesky, P. (2015). Anthropogenic development drives species to be endangered: Capitalism and the decline of species. In A.S. Ragnhild (Ed.), *Green Harms and Crimes: Critical Criminology in a Changing World* (pp. 117–146). Palgrave Macmillan.

Lynch, M. (2003). God's signature: DNA profiling, the new gold standard in forensic science. *Endeavour, 27*(2), 93–97.

Machado, M., & da Silva, I.J.O. (2020). Body expressions of emotions: Do animals have it? *Journal of Animal Behaviour and Biometeorology,* 8(1), 1–10.

MacKenzie, C.A., Dengupta, R.R., & Kaoser R. (2015). Chasing baboons or attending class: Protected areas and childhood education in Uganda. *Environmental Conservation, 42*, 373–383.

Mackenzie, J.S., & Jeggo, M. (2019). The One Health approach – why is it so important? *Tropical Medicine and Infectious Disease, 4*(2), 88. https://doi.org/10.3390/tropicalmed4020088

MacLean, E.L., Snyder-Mackler, N., von Holdt, B.M., & Serpell, J.A. (2019).Highly heritable and functionally relevant breed differences in dog behavior. *Proceedings of the Royal Society B. 286*, 20190716.

MacNamara, M., & MacLean, E. (2017). Selecting animals for education environments. In N.R. Gee, A.H. Fine, & P. McArdle (Eds.), *How Animals Help Students Learn* (pp. 182–196). Routledge.

Madden, F. (2004). Creating coexistence between humans and wildlife: Global perspectives on local efforts to address human-wildlife conflict. *Human Dimensions of Wildlife, 9*, 247–257.

Madden, F., & McQuinn, B. (2014). Conservation's blind spot: The case for conflict transformation in wildlife conservation. *Biological Conservation, 178*, 97–106.

Maharaj, N., & Haney, C.J. (2015). A qualitative investigation of the significance of companion dogs. *Western Journal of Nursing Research, 37*(9), 1175–1193. https://doi.org/10.1177/0193945914545176

Maichomo, M.W., Karanja–Lumumba, T., Olum M.O., Magero, J., Okech, T., & Nyoike, S.N. (2019). *The Status of Donkey Slaughter in Kenya and its Implication on Community Livelihoods*. KALRO. https://kalro.org/sites/default/files/donkey-kalro-report-final-trade-crisis.pdf

Main, M., & Solomon, J. (1986). Discovery of an insecure-disorganized/disoriented attachment pattern. In T.B. Brazelton & M.Q. Yogman (Eds.), *Affective Development in Infancy* (pp. 95–124). Ablex.

Malcolm, R., Ecks, S., & Pickersgill, M. (2017). 'It just opens up their world': Autism, empathy, and the therapeutic effects of equine interactions. *Anthropology & Medicine, 25*(2), 220–234. https://doi.org/10.1080/13648470.2017.1291115

Maller, C.J., Hemsworth, P.H., Ng, K.T., Jongman, E.C., Coleman G.J., & Arnold N.A. (2005). The relationships between characteristics of milking sheds and the attitudes to dairy cows, working conditions, and quality of life of dairy farmers. *Australian Journal of Agricultural Research, 56*, 363–372.

Manciocco, A., Chiarotti, F., & Vitale, A. (2009). Effects of positive interaction with caretakers on the behaviour of socially housed common marmosets (*Callithrix jacchus*). *Applied Animal Behaviour Science, 120*(1–2), 100–107. https://doi.org/10.1016/j.applanim.2009.05.007

Mandler, J.M., Bauer, P.J., & McDonough, L. (1991). Separating the sheep from the goats: Differentiating global categories. *Cognitive Psychology, 23*(2), 263–298. http://dx.doi.org/10.1016/0010-0285(91)9..11-C
Manfredo, M.J. (2008). *Who Cares About Wildlife?* Springer.
Manfredo, M.J., Berl, R.E.W., Teel, T.L., & Bruskotter, J.T. (2021). Bringing social values to wildlife conservation decisions. *Frontiers in Ecology and the Environment,* 19(6), 355–362. https//doi.org/10.1002/fee.2356
Manfredo, M.J., Teel, T.L., & Henry, K.L. (2009). Linking society and environment: A multilevel model of shifting wildlife value orientations in the western United States. *Social Science Quarterly, 90*, 407–427.
Marchant-Forde, J.N, & Boyle, N.A. (2020). COVID-19 effects on livestock production: A One Welfare issue. *Frontiers in Veterinary Science, 7*. https://doi.org/10.3389/fvets.2020.585787
Maréchal, L., Semple, S., Majolo, B., Qarro, M., Heistermann, M., & MacLarnon, A. (2011). Impacts of tourism on anxiety and physiological stress levels in wild male Barbary macaques. *Biological Conservation, 144*(9), 2188–2193.
Marino, L., & Colvin, C.M. (2015). Thinking pigs. *International Journal of Comparative Psychology, 28*(1).
Marino, L., & Lilienfeld, S.O. (2020). Third time's the charm or three strikes you're out? An updated review of the efficacy of dolphin-assisted therapy for autism and developmental disabilities. *Journal of Clinical Psychology, 1*, 15. https://doi.org/10.1002/jclp.23110
Marker, L.L., Dickman, A.J., & Macdonald, D.W. (2005). Perceived effectiveness of livestock-guarding dogs placed on Namibian farms. *Rangeland Ecology & Management, 58*, 329–336.
Markman, E.M., & Abelev, M. (2004). Word learning in dogs? *Trends in Cognitive Sciences, 8*, 479–481.
Marshall-Pescini, S., Cafazzo, S., Virányi, Z., & Range, F. (2017). Integrating social ecology in explanations of wolf–dog behavioral differences. *Current Opinion in Behavioural Sciences, 16*, 80–86.
Martens, P., Enders-Slegers, M.J., & Walker, J.K. (2016). The emotional lives of companion animals: Attachment and subjective claims by owners of cats and dogs. *Anthrozoös, 29*(1), 73–88.
Martin, F., & Farnum, J. (2002). Animal-assisted therapy for children with pervasive developmental disorders. *Western Journal of Nursing Research, 24*(6), 657–670.
Martin, P., & Shepheard, M. (2011). What is meant by the social license? In J. Williams & P. Martin(Eds.), *Defending the Social License of Farming: Issues, Challenges and New Directions for Agriculture* (pp. 3–11). CSIRO.
Marvin, G. (1984). The cockfight in Andalusia (Spain): Images of the truly male. *Anthropological Quarterly, 57*(2), 60–70.
Marvin, G. (2012). *Wolf.* Reaktion Books.
Matos, R.E., Jakuba, T., Mino, I., Fejsakova, M., Demeova, A., & Kottferova, J. (2015). Characteristics and risk factors of dog aggression in the Slovak Republic. *Veterinární Medicina, 60*(8), 432–445.
Matsumoto, S., Iwadate, K., Aoyagi, M., Ochiai, E., Ozawa, M., & Asakura, K. (2013). An experimental study on the macroscopic findings of ligature marks using a murine model. *American Journal of Forensic Medicine and Pathology, 34*(1), 72–74.
Maust-Mohl, M., Fraser, J., & Morrison, R. 2012. Wild minds: What people think about animal thinking. *Anthrozoös, 25*(2), 133–147.
Maynard, L., Jacobson, S.K., Monroe, M.C., & Savage, A. (2019). Mission impossible or mission accomplished: Do zoo organizational missions influence conservation practice? *Zoo Biology, 39*, 304–314.
McConnell, A.R., Brown, C.M., Shoda, T.M., Stayton, L.E., & Matrin, C.E. (2011). Friends with benefits: On the positive consequences of pet ownership. *Journal of Personality and Social Psychology, 101*(6), 1239–1252.
McCulloch, S.P., & Reiss, M.J. (2017). Bovine tuberculosis and badger control in Britain: Science, policy and politics. *Journal of Agricultural and Environmental Ethics, 30*(4), 469–484.
McCune, S., Esposito, L., & Griffin, J.A. (2017). Introduction to the thematic series on animal assisted interventions in special populations. *Applied Developmental Science, 21*(2), 136–138.
McCune, S., McCradle, P., Griffin, J.A., Esposito, L., Hurley, K., Bures, R. et al. (2020). Human-animal interaction (HAI) research: A decade of progress. *Frontiers in Veterinary Science, 7*, 44. https://doi.org/10.3389/fvets.2020.00044.

McGreevy, P., & Boakes, R. (2011). *Carrots and Sticks*. Darlington Press.
McNicholas, J., & Collis, G.M. (2000). Dogs as catalysts for social interactions: Robustness of the effect. *British Journal of Psychology, 91*(1), 61–70. https://doi.org/10.1348/000712600161673
McNicholas, J., & Collis, G.M. (2006). Animals as social supports: Insights for understanding animal-assisted therapy. In A.H. Fine (Ed.), *Handbook on Animal-Assisted Therapy* (pp. 49–59). Elsevier.
Meaney, M.J., Szyf, M. (2005). Maternal care as a model for experience-dependent chromatin plasticity? *Trends in Neurosciences, 28*(9), 456–463.
Meehan M.K., & Shackelford T.K. (2021). Neoteny. In J. Vonk & T. Shackelford (Eds.), *Encyclopedia of Animal Cognition and Behavior* (pp. 1–3). Springer. https://doi.org/10.1007/978-3-319-47829-6_557-1
Meehan, M., Massavelli, B., & Pachana, N. (2017). Using attachment theory and social support theory to examine and measure pets as sources of social support and attachment figures. *Anthrozoös, 30*(2), 273–289.
Meints, K., Brelsford, V., & De Keuster, T. (2018). Teaching children and parents to understand dog signaling. *Frontiers in Veterinary Science, 20*.
Meints, K., Brelsford, V., Gee, N.R., & Fine, A. (2017). Animals in educational settings – safety for all. In A. Fine & N. Gee (Eds.), *How Animals Help Students Learn: Research and Practice for Educators and Mental-Health Professionals* (pp. 12–26). Routledge.
Meints, K., Brelsford, V.L., Dimolareva, M., Maréchal, L., Pennington, K., Rowan, E. et al. (2022). Can dogs reduce stress levels in school children? Effects of dog-assisted interventions on salivary cortisol in children with and without special educational needs using randomized controlled trials. *PLOS ONE, 17*(6), e0269333.
Melfi, V.A., & Thomas, S. (2005). Can training zoo-housed primates compromise their conservation? A case study using Abyssinian colobus monkeys (*Colobus guereza*). *Anthrozoös, 18*(3), 304–317. https://doi.org/10.2752/089279305785594063
Melson, G.F. (2003). Child development and the human-companion animal bond. *American Behavioral Scientist, 47*(1), 31–39. https://doi.org/10.1177/0002764203255210
Melson, G.F., & Fine, A.H. (2015). Animals in the lives of children. In A.H. Fine (Ed.), *Animal-Assisted Therapy: Foundations and Guidelines for Animal-Assisted Interventions* (4th ed., pp. 179–194). Academic Press.
Menezes, R. (2008). Rabies in India. *Canadian Medical Association Journal, 178*, 564–566.
Menor-Campos, D.J., Hawkins, R., & Williams, J. (2018). Belief in animal mind among Spanish primary school children. *Anthrozoös, 31*(5), 599–614.
Menotti-Raymond, M.A., David, V.A., & O'Brien, S.J. (1997). Pet cat hair implicates murder suspect. *Nature, 386*(6627), 774–774.
Mepham, B. (1997). Ethical analysis of food biotechnologies: An evaluative framework. In B. Mepham (Ed.), *Food Ethics* (pp. 101–119). Routledge.
Mepham, B. (2000). 'Wurde der Kreature' and the common morality. *Journal of Agricultural and Environmental Ethics, 13*, 65–78.
Mepham, B. (2008a). *Bioethics: An Introduction for the Biosciences* (2nd ed.). Oxford University Press.
Mepham, B. (2008b). A notional ethical contract with farm animals in a sustainable global food system. In R. Fish, S. Seymour, M. Steven, C. Watkins (Eds.), *Sustainable Farmland Management: Transdisciplinary Approaches* (pp. 125–134). CABI.
Mepham, B. (2012a). Agricultural ethics. In R. Chadwick (Ed.), *Encyclopedia of Applied Ethics* (2nd ed., Vol. 1, pp. 86–96). Academic Press.
Mepham, B. (2012b). Food ethics. In R. Chadwick (Ed.), *Encyclopedia of Applied Ethics* (2nd ed., Vol. 2, pp. 322–330). Academic Press.
Mepham, B. (2013). Ethical principles and the ethical matrix. In J.P. Clark and C. Ritson (Eds.), *Practical Ethics for Food Scientists* (pp. 39–56). Wiley-Blackwell.
Mepham, B. (2016). Morality, morbidity and mortality: An ethical analysis for culling nonhuman animals. In L.B. Meijboom & E.N. Stassen (Eds.), *The End of Animal life: A Start for Ethical Debate* (pp. 117–135). Wageningen Academic Publishers.
Mepham, T.B. (1991). Bovine somatotropin and public health. *British Medical Journal, 2*(6775), 483.

Mepham, T.B. (1992). Public health implications of bovine somatotropin use in dairying: Discussion paper. *Journal of the Royal Society of Medicine, 85*(12), 736.
Merton, R. (1968). *Delinquency & Drift*. Transaction.
Mikulincer, M., Shaver, P.R., & Pereg, D. (2003). Attachment theory and affect regulation: The dynamics, development, and cognitive consequences of attachment-related strategies. *Motivation and Emotion, 27*(2), 77–102.
Miletski, H. (2002). *Understanding Bestiality and Zoophilia*. East-West Publishing.
Miller, S.C., Kennedy, C.C., DeVoe, D.C., Hickey, M., Nelson, T., & Kogan, L. (2009). An examination of changes in oxytocin levels in men and women before and after interaction with a bonded dog. *Anthrozoös, 22*(1), 31–42.
Mitchell, R.W. (1997). Kinesthetic-visual matching and the self-concept as explanations of mirror-self-recognition. *Journal for the Theory of Social Behaviour, 27*(1), 17–39.
Mitchell, R.W. (2012). Self-recognition in animals. In M.R. Leary & J.P. Tangney (Eds.), *Handbook of Self and Identity* (pp. 656–679). Guilford Press.
Mogomotsi, P.K., Mogomotsi, G.E.J., Dipogiso, K., Phonchi-Tshekiso, N.D., Stone, L.S., & Badimo, D. (2020). An analysis of communities' attitudes toward wildlife and implications for wildlife sustainability. *Tropical Conservation Science, 13*. https://doi.org/10.1177/1940082920915603
Molloy, C. (2011). *Popular Media and Animals*. Palgrave Macmillan.
Moorhouse, T., D'Cruze, N.C., & Macdonald, D.W. (2016). Unethical use of wildlife in tourism: What's the problem, who is responsible, and what can be done?. *Journal of Sustainable Tourism, 25*(4), 505–516.
Morand, S. (2020). Emerging diseases, livestock expansion and biodiversity loss are positively related at global scale. *Biological Conservation, 248*, 108707.
Morey, D., & Wiant, M.D. (1992), Early holocene domestic dog burials from the North American Midwest. *Current Anthropology, 33*(2), 224–229.
Morris, P.H., Doe, C., & Godsell, E. (2008). Secondary emotions in non-primate species? Behavioural reports and subjective claims by animal owners. *Cognition and Emotion, 22*(1), 3–20.
Moscardo, G., Woods, B., & Saltzer, R. (2004). The role of interpretation in wildlife tourism. In K. Higginbottom (Ed.), *Wildlife Tourism: Impacts, Management and Planning* (pp. 231–251). Common Ground.
Mota-Rojas, D., Mariti, C., Zdeinert, A., Riggio, G., Mora-Medina, P., del Mar Reyes, A. et al. (2021). Anthropomorphism and its adverse effects on the distress and welfare of companion animals. *Animals, 11*(11), 3263.
Mota-Rojas, D., Monsalve, S., Lezama-García, K., Mora-Medina, P., Domínguez-Oliva, A., Ramírez-Necoechea, R. et al. (2022). Animal abuse as an indicator of domestic violence: One Health, One Welfare approach. *Animals, 12*(8), 977.
Mubanga, M., Byberg, L., Nowak, C., Egenvall, A., Magnusson, P.K., Ingelsson, E. et al. (2017). Dog ownership and the risk of cardiovascular disease and death – a nationwide cohort study. *Scientific Reports, 7*(1), 1–9.
Muldoon, J.C., Williams, J.M., Lawrence, A., & Currie, C. (2019). The nature and psychological impact of child/adolescent attachment to dogs compared with other companion animals. *Society & Animals, 27*(1), 55–74.
Müller, C.A., Schmitt, K., Barber, A.L.A., & Huber, L. (2015). Dogs can discriminate emotional expressions of human faces. *Current Biology, 25*, 601–605. https://doi.org/10.1016/ j.cub.2014.12.05
Murphy, R., & Daly, S.L. (2020). Psychological distress among non-human animal rescue workers: an exploratory study. *Society & Animals, 1*, 1–21 (published online ahead of print). https://doi.org/10.1163/15685306-BJA10028
Mutanga, C.N., Chikuta, O., Muboko, N., Gandiwa, E., Kabote, F., & Kaswaurere, T.W. (2020). Wildlife tourism, conservation and community benefits in Zimbabwe. In M. Novelli, E.A. Adu-Ampong, & M.A. Ribeiro (Eds.), *Routledge Handbook of Tourism in Africa* (pp. 459–473). Routledge.

Nagai, T., Aoyagi, M., Ochiai, E., Sakai, K., Maruyama-Maebashi, K., Fukui, K. et al. (2011). Longitudinal evaluation of immunohistochemical findings of milk aspiration: An experimental study using a murine model. *Forensic Science International, 209*(1–3), 183–185.

Nagasawa, M., Kikusui, T., Onaka, T., & Ohta, M. (2009). Dog's gaze at its owner increases owner's urinary oxytocin during social interaction. *Hormones and Behavior, 55*(3), 434–441.

Nagasawa, M., Mitsui, S., En, S., Ohtani, N., Ohta, M., Sakuma, Y. et al. (2015). Oxytocin-gaze positive loop and the coevolution of human-dog bonds. *Science, 348*(6232), 333–336.

Nagasawa, M., Murai, K., Mogi, K., & Kikusui, T. (2011). Dogs can discriminate human smiling faces from blank expressions. *Animal Cognition, 14*, 525–533. https://doi.org/ 10.1007/s10071-011-0386-5

Nakajima, S. (2015). Dogs and owners resemble each other in the eye region. *Anthrozoös, 26*(4), 551–556. https://doi.org/10.2752/175303713X13795775536093

Nakajima, S., Yamamoto, M., & Yoshimoto, N. (2015). Dogs look like their owners: Replications with racially homogenous owner portraits. *Anthrozoös, 22*(2), 173–181. https://doi.org/10.2752/175303709 X434194

Naughton-Treves, L. (1997). Farming the forest edge: Vulnerable places and people around Kibale National Park, Uganda. *Geographical Review, 87*, 27–46.

Newsome, D., Dowling, R., & Moore, S. (2005). *Wildlife Tourism*. Channel View.

Ng, V.C., Lit, A.C., Wong, O.F., Tse, M.L., & Fung, H.T. (2018). Injuries and envenomation by exotic pets in Hong Kong. *Hong Kong Medical Journal, 24*(1), 48–55.

Nielsen, M., Suddendorf, T., & Slaughter, V. (2006). Mirror self-recognition beyond the face. *Child Development, 77*(1), 176–185.

Nijland, M.L., Stam, F., & Seidell, J.C. (2010). Overweight in dogs, but not in cats, is related to overweight in their owners. *Public Health Nutrition, 13*(1), 102–106. https://doi.org/10.1017/S136898000999022X

Nijman, V., Smith, J.H., Foreman, G., Campera, M., Feddema, K., & Nekaris, K.A.I. (2021). Monitoring the trade of legally protected wildlife on Facebook and Instagram illustrated by the advertising and sale of apes in Indonesia. *Diversity, 13*(6), 236.

Nimmo, R. (2012). Animal cultures, subjectivity and knowledge: Symmetrical reflections beyond the great divide. *Society and Animals, 20*, 173–192.

Nóbrega Alves, R.R., & Duarte Barboza, R.R. (2018). From Roman arenas to movie screens: Animals in entertainment and sport. In R.R. Nóbrega Alves & U.P. Albuquerque (Eds.), *Ethnozoology: Animals in Our Lives* (pp. 363–382). Academic Press.

Nurse, A. (2011). Policing wildlife: Perspectives on criminality in wildlife crime. *Papers from the British Criminology Conference, 11*, 38–53.

Nurse, A. (2012). Repainting the thin green line: The enforcement of UK wildlife law. *Internet Journal of Criminology, 12*, 1–20.

Nurse, A. (2013). *Animal Harm: Perspectives on Why People Harm and Kill Animals*. Ashgate.

Nurse, A., & Wyatt, T. (2020). *Wildlife Criminology*. Bristol University Press.

O'Haire, M.E., McKenzie, S.J., Beck, A.M., & Slaughter, V. (2015). Animals may act as social buffers: Skin conductance arousal in children with autism spectrum disorder in a social context. *Developmental Psychobiology, 57*(5), 584–595. https://doi.org/10.1002/dev.21310

O'Haire, M.E., McKenzie, S.J., McCune, S., & Slaughter, V. (2013). Effects of classroom animal-assisted activities with guinea pigs in the primary school classroom. *Anthrozoös, 26*(3), 445–458. https://doi.org/ 10.2752/175303713X13697429463835

O'Neill, O. (1993). Kantian ethics. In P. Singer (Ed.), *A Companion to Ethics* (pp. 175–185). Blackwell.

Odendaal, J.S. (2000). Animal-assisted therapy—magic or medicine? *Journal of Psychosomatic Research, 49*(4), 275–280.

Odendaal, J.S.J., & Meintjes, R.A., (2003). Neurophysiological correlates of affiliative behaviour between humans and dogs. *Veterinary Journal, 165*, 296–301. https://doi.org/10.1016/s1090-0233(02)00237-X.

Ogra, M.V. (2008). Human-wildlife conflict and gender in protected area borderlands: A case study of costs, perceptions, and vulnerabilities from Uttarakhand (Uttaranchal), India. *Geoforum, 39*, 1408–1422.

Oliva, J.L., Wong, Y.T., Rault, J.L., Appleton, B., & Lill, A. (2016).The oxytocin receptor gene, an integral piece of the evolution of Canis familaris from Canis lupus. *Pet Behaviour Sci*ence*, 1*, 1–15.

Olsson, I.A.S., Nielsen, B.L., Camerlink, I., Pongrácz, P., Golledge, H.D., Chou, J.Y. et al. (2022). An international perspective on ethics approval in animal behaviour and welfare research. *Applied Animal Behaviour Science, 253*, 105658.

Ontillera-Sánchez, R.R. (2020). *Of Casteadores,Gallos y Galleras: The Cockfight World in the Canary Islands*. [Unpublished doctoral thesis]. University of Roehampton (London).

Orams, M.B. (1996). A conceptual model of tourist-wildlife interaction: The case for education as a management strategy. *Australian Geographer, 27*(1), 39–51.

Orams, M.B. (2002). Feeding wildlife as a tourism attraction: A review of issues and impacts. *Tourism Management, 23*(3), 281–293.

Organization for Economic Cooperation and Development (OECD). (2022). https://data.oecd.org/agroutput/meat-consumption.htm

Ostrander, E.A., Wang, G.D., Larson, G., Vonholdt, B.M., Davis, B.W., Jagannathan, V. et al. (2019). Dog10K: An international sequencing effort to advance studies of canine domestication, phenotypes and health. *National Science Review, 6*(4), 810–824.

Overall, K.L., & Love, M. (2001). Dog bites to humans – demography, epidemiology, injury, and risk. *Journal of the American Veterinary Medical Association, 218*(12), 1923–1934.

Ovodov, N.D., Crockford, S.J., Kuzmin, Y.V., Higham, T.F.G., Hodgins, G.W.L., & van der Pflicht, J. (2011). A 33,000-year-old incipient dog from the Altai Mountains of Siberia: Evidence of the earliest domestication disrupted by the Last Glacial Maximum, *PLOS ONE, 6*(7), e22821.

Owen-Smith, N.G.I.H., Kerley, G.I.H., Page, B., Slotow, R., & Van Aarde, R.J. (2006). A scientific perspective on the management of elephants in the Kruger National Park and elsewhere: Elephant conservation. *South African Journal of Science, 102*(9), 389–394.

Ownby, D.R., Johnson, C.C., & Peterson, E.L. (2002). Exposure to dogs and cats in the first year of life and risk of allergic sensitization at 6 to 7 years of age. *JAMA, 288*(8), 963–972.

Oxley J.C., & Waggoner, L.P. (2011). Detection of explosive by dogs. In M.M. Marshall & J.C. Oxley (Eds.), *Aspects of Explosives Detection* (pp. 27–40). Elsevier.

Pagel, C., Orams, M., & Lück, M. (2020). #BiteMe: Considering the potential influence of social media on in-water encounters with marine wildlife. *Tourism in Marine Environments, 15*(3–4), 249–258.

Panksepp, J. (2005). Affective consciousness: Core emotional feelings in animals and humans. *Consciousness and Cognition, 14*, 30–80, 2005.

Parfitt, C.H., & Alleyne, E. (2018). Animal abuse proclivity: Behavioral, personality and regulatory factors associated with varying levels of severity. *Psychology, Crime & Law, 24*(5), 538–557. https://doi.org/10.1080/1068316X.2017.1332193

Park, H., Chun, M.S., & Joo, Y. (2020). Traumatic stress of frontline workers in culling livestock animals in South Korea. *Animals, 10*(10), 1920. https://doi.org/10.3390/ani10101920

Parkin, T.D.H., Brown, J., & Macdonald, E.B. (2018). Occupational risks of working with horses: A questionnaire survey of equine veterinary surgeons. *Equine Veterinary Education, 30*, 200–205. https://doi.org/10.1111/eve.12891

Pedersen, I., Martinsen, E.W., Berget, B., & Braastad, B.O. (2012). Farm animal-assisted intervention for people with clinical depression: A randomized controlled trial. *Anthrozoös, 25*(2), 149–160. https://doi.org/10.2752/175303712X13316289505260

Pedersen, V., Barnett, J.L., Hemsworth, P.H., Newman, E.A., & Schirmer, B. (1998). The effects of handling on behavioural and physiological responses to housing in tether-stalls in pregnant pigs. *Animal Welfare, 7*, 137–150.

Pendleton, A.L., Shen, F., Taravella, A.M., Emery, S., Veeramah, K.R., Boyko, A.R., & Kidd, J.M. (2018). Comparison of village dog and wolf genomes highlights the role of the neural crest in dog domestication. BMC Biology, 16(1), 1–21.

Pepperberg, I.M. (2002). Cognitive and communicative abilities of grey parrots. *Current Directions in Psychological Science, 11*(3), 83–87.

Pepperberg, I.M., Gardiner, L.I., & Luttrell, L.J. (1999). Limited contextual vocal learning in the grey parrot (*Psittacus erithacus*): The effect of interactive co-viewers on videotaped instruction. *Journal of Comparative Psychology, 113*(2), 158.

Pereira, J., & Fonte, D. (2018). Pets enhance antidepressant pharmacotherapy effects in patients with treatment resistant major depressive disorder. *Journal of Psychiatric Research, 104*, 108–113. https://doi.org/10.1016/j.jpsychires.2018.07.004

Perret, J.L., Best, C.O., Coe, J.B., Greer, A.L., Khosa, D.K., & Jones-Bitton, A. (2020). Prevalence of mental health outcomes among Canadian veterinarians. *Journal of the American Veterinary Medical Association, 256*(3), 365–375. https://doi.org/10.2460/javma.256.3.365

Pet Food Manufacturer's Association (PFMA). (2021). Pet Population 2021. www.pfma.org.uk/historical-pet-population

Peterson, M.N., Birckhead, J.L., Leong, K., Peterson, M.J., Peterson, T.R. et al. (2010). Rearticulating the myth of human-wildlife conflict. *Conservation Letters, 3*, 74–82.

Petkov, C.I., Flecknell, P., Murphy, K., Basso, M.A., Mitchell, A., Hartig, R. et al. (2022). Unified ethical principles and an animal research 'Helsinki' declaration as foundations for international collaboration. *Current Research in Neurobiology, 3*, 100060.

Philpotts, I., Dillon, J., & Rooney, N. (2019). Improving the welfare of companion dogs – Is owner education the solution? *Animals, 9*(9), 662.

Pilley, J.W., & Hinzmann, H. (2014). *Chaser: Unlocking the Genius of the Dog Who Knows 1000 Words*. Simon and Schuster.

Pilley, J.W., & Reid, A.K. (2011). Border collie comprehends object names as verbal referents. *Behavioural Processes, 86*(2), 184–195.

Pinillos, R.G. (2018). *One Welfare: A Framework to Improve Animal Welfare and Human Well-Being*. Cab International.

Pinillos, R.G., Appleby, M.C., Manteca, X., Scott-Park, F., Smith, C., & Velarde, A. (2016). One Welfare – a platform for improving human and animal welfare. *Veterinary Record, 179*(16), 412–413.

Pitt, J. (2017). *The Ecology of Chickens: An Examination of the Introduction of the Domestic Chicken Across Europe After the Bronze Age* [Unpublished doctoral thesis]. Bournemouth University.

Plotnik, J.M., de Waal, F.B., & Reiss, D. (2006). Self-recognition in an Asian elephant. *Proceedings of the National Academy of Sciences, 103*(45), 17053–17057.

Podberscek, A. (2009). Good to pet and eat: The keeping and consuming of dogs and cats in South Korea. *Journal of Social Issues, 65*(3), 615–632. https://doi.org/10.1111/j.1540-4560.2009.01616.x

Polachek, A.J., & Wallace, J.E. (2018). The paradox of compassionate work: A mixed-methods study of satisfying and fatiguing experiences of animal health care providers. Anxiety, Stress, & Coping, 31(2), 228–243.

Polheber, J.P., & Matchock, R.L. (2014). The presence of a dog attenuates cortisol and heart rate in the trier social stress test compared to human friends. *Journal of Behavioral Medicine, 37*(5), 860–867.

Pollux, P.M., Craddock, M., & Guo, K. (2019). Gaze patterns in viewing static and dynamic body expressions. *Acta Psychologica, 198*, 102862.

Pooley, S. (2016). A cultural herpetology of Nile crocodiles in Africa. *Conservation and Society, 14*(4), 391–405.

Pooley, S., Bhatia, S., & Vasava, A. (2021). Rethinking the study of human–wildlife coexistence. *Conservation Biology, 35*(3), 784–793.

Pörtl, D., & Jung, C. (2017). Is dog domestication due to epigenetic modulation in brain? *Dog Behavior, 3*(2), 21–32.

Pörtl, D., & Jung, C. (2019). Physiological pathways to rapid prosocial evolution. *Biologia Futura, 70*, 93–102.

Potter, K., Teng, J.E., Masteller, B., Rajala, C., & Balzer, L.B. (2019). Examining how dog 'acquisition' affects physical activity and psychosocial well-being: Findings from the BuddyStudy pilot trial. *Animals, 9*(9), 666. https://doi.org/10.3390/ANI9090666

Powell, L., Guastella, A.J., McGreevy, P., Bauman, A., Edwards, K.M., & Stamatakis, E. (2019). The physiological function of oxytocin in humans and its acute response to human-dog interactions: A review of the literature. *Journal of Veterinary Behavior, 30*, 25–32. https://doi.org/10.1016/j.jveb.2018.10.008

Powys, J.C. (1974). *The Art of Growing Old*. Village Press.

Prat, N., Rongieras, F., de Freminville, H., Magnan, P., Debord, E., Fusai, T. et al. (2012). Comparison of thoracic wall behavior in large animals and human cadavers submitted to an identical ballistic blunt thoracic trauma. *Forensic Science International, 222*(1–3), 179–185.

Pręgowski, P.M., & Cieślik, S. (2020). Attitudes to animal abuse in veterinary practice in Poland, *Anthrozoös, 33*(3), 427–440, https://doi.org/10.1080/08927936.2020.1746532

Price, T., Wadewitz, P., Cheney, D., Seyfarth, R., Hammerschmidt, K., & Fischer, J. (2015). Vervets revisited: A quantitative analysis of alarm call structure and context specificity. *Scientific Reports, 5*(1), 1–11.

Prior, H., Schwarz, A., & Güntürkün, O. (2008). Mirror-induced behavior in the magpie (*Pica pica*): Evidence of self-recognition. *PLOS Biology, 6*(8), e202.

Probyn-Rapsey, F. (2018). Anthropocentrism. In L. Gruen (Ed.), *Critical Terms for Animal Studies* (pp. 47–63). University of Chicago Press.

Prothmann, A., Ettrich, C., & Prothmann, S. (2009). Preference for, and responsiveness to, people, dogs and objects in children with autism. *Anthrozoös, 22*(2), 161–171.

Purewal, R., Christley, R., Kordas, K., Joinson, C., Meints, K., Gee, N. et al. (2017). Companion animals and child/adolescent development: A systematic review of the evidence. *International Journal of Environmental Research and Public Health, 14*(3), 234.

Quesque, F., & Rossetti, Y. (2020). What do theory-of-mind tasks actually measure? Theory and practice. *Perspectives on Psychological Science, 15*(2), 384–396.

Racca, A., Amadei, E., Ligout, S., Guo, K., Meints, K., & Mills, D. (2010). Discrimination of human and dog faces and inversion responses in domestic dogs (*Canis familiaris*). *Animal Cognition, 13*, 525–533. https://doi.org/10.1007/s10071-009-0303-3

Racca, A., Guo, K., Meints, K., & Mills, D.S. (2012). Reading faces: Differential lateral gaze bias in processing canine and human facial expressions in dogs and 4-year-old children. *PLOS ONE, 7*, 36076. https://doi.org/10.1371/journal.pone.0036076

Radford, M. (2001). *Animal Welfare Law in Britain: Regulation and Responsibility*. Oxford University Press.

Ramasawmy, M., Galiè, A., & Dessie, T. (2018). *Poultry Trait Preferences and Gender in Ethiopia*. CGIAR Gender and Breeding Initiative Working Paper.

Ramasawmy, M.R. (2017). *Do 'chickens dream only of grain'? Uncovering the Social Role of Poultry in Ethiopia*. [Unpublished doctoral thesis]. University of Roehampton (London).

Ramos, D., & Ades, C. (2012). Two-item sentence comprehension by a dog (*Canis familiaris*). *PLOS ONE, 7*(2), e29689.

Range, F., & Virányi, Z. (2011). Development of gaze following abilities in wolves (*Canis lupus*). *PLOS ONE, 6*(2), e16888.

Rault, J.L., Waiblinger, S., Boivin, X., & Hemsworth, P. (2020). The power of a positive human–animal relationship for animal welfare. *Frontiers in Veterinary Science, 7*(November), 1–13. https://doi.org/10.3389/fvets.2020.590867

Raupp, C.D. (1999). Treasuring, trashing or terrorizing: Adult outcomes of childhood socialization about companion animals. *Society & Animals, 7*, 141–159. https://doi.org/10.1163/156853099X00040

Ravel, A., D'Allaire, S., Bigras-Poulin, M., & Ward, R. (1996). Influence of management, housing and personality of the stockperson on preweaning performances on independent and integrated swine farms in Quebec. *Preventive Veterinary Medicine, 29*, 37–57.

Ravenscroft, S.J., Barcelos, A.M., & Mills, D.S. (2021). Cat-human related activities associated with human well-being. *Human-Animal Interaction Bulletin, 11*(2), 79–95. https://doi.org/10.1079/hai.2021.0006

Rawls, J. (1951). Outline for a decision procedure for ethics. *Philosophical Review, 60*, 177–197.

Rawls, J. (1971). *A Theory of Justice*. Harvard University Press.

Redpath, S.M., Young, J., Evely, A., Adams, W.M., Sutherland, W.J., Whitehouse, A. et al. (2013). Understanding and managing conservation conflicts. *Trends in Ecology and Evolution, 28*, 100–109. https//doi.org/10.1016/j.tree.2012.08.021

Reefmann, N., Wechsler, B., & Gygax, L. (2009). Behavioural and physiological assessment of positive and negative emotion in sheep. *Animal Behaviour, 78*(3), 651–659. https://doi.org/10.1016/j.anbehav.2009.06.015

Regan, T. (2006). Sentience and rights. In J. Turner & J. D'Silva (Eds.), *Animals, Ethics and Trade* (pp. 79–86). Earthscan.

Reimert, I., Fong, S., Rodenburg, T.B., & Bolhuis, J.E. (2017). Emotional states and emotional contagion in pigs after exposure to a positive and negative treatment. *Applied Animal Behaviour Science, 193*, 37–42.

Reinisch, A.I. (2008). Understanding the human aspects of animal hoarding. *Canadian Veterinary Journal, 49*(12), 1211. http://pubmedcentralcanada.ca/pmcc/articles/PMC2583418/

Reiser, D. (2017). Will the ark sink? Captive wildlife, tourism and the human relationship to nature: Demystifying zoos. In I. Borges de Lima & R.J. Green (Eds.), *Wildlife Tourism, Environmental Learning, and Ethical Encounters* (pp. 263–272). Springer.

Reiss, D., & Marino, L. (2001). Mirror self-recognition in the bottlenose dolphin: A case of cognitive convergence. *Proceedings of the National Academy of Sciences, 98*(10), 5937–5942.

Reynolds, P.C., & Braithwaite, D. (2001). Towards a conceptual framework for wildlife tourism. *Tourism Management, 22*, 31–42.

Rice, M., Hemsworth, L., Hemsworth, P.H., & Coleman, G. (2020). The impact of a negative media event on public attitudes towards animal welfare in the red meat industry. *Animals, 10*, 619.

Roberts, B.W. (2009). Back to the future: Personality and assessment and personality development. *Journal of Research in Personality, 43*(2), 137–145. https://doi.org/10.1016/J.JRP.2008.12.015

Roberts, M., Cook, D., Jones, P., & Lowther, D. (2001). *Wildlife Crime in the UK: Towards a National Crime Unit.* Department for Environment Food & Rural Affairs/Centre for Applied Social Research (University of Wolverhampton).

Rock, R.C., Haugh, S., Davis, K.C., Anderson, J.L., Johnson, A.K., Jones, M.A. et al. (2021). Predicting animal abuse behaviors with externalizing and psychopathic personality traits. *Personality and Individual Differences, 171*, 110444. https://doi.org/10.1016/j.paid.2020.110444

Rodrigues, J.B., Schlechter, P., Spychiger, H., Spinelli, R., Oliveira, N., & Figueiredo T. (2017). The XXI century mountains: Sustainable management of mountainous areas based on animal traction. *Open Agriculture, 2*(1), 300–7.

Rodriguez, K.E., Herzog, H., & Gee, N.R. (2021). Variability in human-animal interaction research. *Frontiers in Veterinary Science, 7*, 619600. https://doi.org/10.3389/fvets.2020.619600

Rollin, B.E. (2006). *Animal Rights and Human Morality*. Prometheus Books.

Root-Gutteridge, H., Ratcliffe, V.F., Korzeniowska, A.T., & Reby, D. (2019). Dogs perceive and spontaneously normalize formant-related speaker and vowel differences in human speech sounds. *Biology Letters, 15*, 20190555. https://doi.org/10.1098/rsbl.2019.0555

Ross, W.D. (1930). *The Right and the Good.* Clarendon Press.

Rossano, F. (2018). Social manipulation, turn-taking and cooperation in apes: Implications for the evolution of language-based interaction in humans. *Interaction Studies, 19*(1–2), 151–166.

Roth, L.S.V., Faresjö, Å., & Theodorsson, E. (2015). Hair cortisol varies with season and lifestyle and relates to human interactions in German shepherd dogs. *Scientific Reports, 6*, 19631. https://doi.org/10.1038/srep19631

Rouha-Mulleder, C., Iben, C., Wagner, E., Laaha, G., Troxler, J., & Waiblinger, S. (2009). Relative importance of factors influencing the prevalence of lameness in Austrian cubicle loose-housed dairy cows. *Preventive Veterinary Medicine, 92*, 123–133.

Roy, M.M., & Christenfeld, N.J.S. (2004). Do dogs resemble their owners? *Psychological Science, 15*(5), 363.

Roy, M.M., & Christenfeld, N.J.S. (2005). Dogs still do resemble their owners. *Psychological Science, 16*(9), 743–744.

Royal Society for the Prevention of Cruelty to Animals. (2013). *Prosecutions Department Annual Report 2013*. www.rspca.org.uk/ImageLocator/LocateAsset?asset¼ document&assetId¼1232735296611&mode¼prd

Ruiz, R., Díez-Unquera, B., De Heredia, I.B., Mandaluniz, N., Arranz, J., & Ugarte, E. (2010). The Latxa dairy sheep in the Basque Country: Importance, challenges and opportunities for a traditional livestock activity. In C.M. Romeo Casabona, L. Escajedo San Epifanio, & A.E. Cirión (Eds.), Global *F*ood *S*ecurity: Ethical and *L*egal *C*hallenges (pp. 138–140). Wageningen Academic Publishers.

Rujoiu, O., & Rujoiu, V. (2013). Human-animal bond: Loss and grief. A review of the literature. *Revista de Asistenta Sociala, 3*, 163–171. www.researchgate.net/publication/287995122

Rushton, J., & Bruce, M. (2017). Using a one health approach to assess the impact of parasitic disease in livestock: How does it add value? *Parasitology, 144*(1), 15–25.

Russow, L.M. (2002). Ethical implications of the human-animal bond in the laboratory. *Ilar Journal, 43*(1), 33–37.

Rynearson, E. (1978). Humans and pet attachment. *Psychiatry, 133*, 550–555

Sable, P. (2013). The pet connection: An attachment perspective. *Clinical Social Work Journal, 41*(1), 93–99.

Saetre, P., Lindberg, J., Leonard, J.A., Olsson, K., Pettersson, U., Ellegren, H. et al. (2004). From wild wolf to domestic dog: Gene expression changes in the brain. *Molecular Brain Res*earch, *126*, 198–206.

Sakagami, T., & Ohta, M. (2010). The effect of visiting zoos on human health and quality of life. *Animal Science Journal, 81*(1), 129–134. https://doi.org/10.1111/J.1740-0929.2009.00714.X

Salomons, H., Smith, K., Callahan-Beckel, M., Callahan, M., Levy, K., Kennedy, B.S. et al. (2021).Cooperative Communication with Humans Evolved to Emerge early in dogs. *Current Biology, 31* (14), 3137–3144. doi.org.{10.1016\j.cub.2021.06.051

Sánchez-Villagra, M.R., Geiger, M., & Schneider, R.A. (2016). The taming of the neural crest: A developmental perspective on the origins of morphological covariation in domesticated mammals. *Royal Society Open Science, 3*(6), 160107.

Sandbrook, C., & Adams, W.M. (2012). Accessing the impenetrable: The nature and distribution of tourism benefits at a Ugandan national park. *Society & Natural Resources, 25*(9), 915–932.

Sanders, C.E., & Henry, B.C. (2018). The role of beliefs about aggression in cyberbullying and animal abuse. *Psychology, Crime & Law, 24*(5), 558–571. https://doi.org/10.1080/1068316X.2017.1327585

Sanders, C.R., Rasmussen, J.L., Modlin, S.J., Holder, A.M., & Rajecki, D.W. (1999). Good dog: Aspects of humans' causal attributions for a companion animal's social behavior. *Society & Animals, 7*(1), 17–34.

Sandøe, P. (2008). Re-thinking the ethics of intensification for animal agriculture: Comments on David Fraser, Animal Welfare and the Intensification of Animal Production. In P.B. Thompson (Ed.), *The* E*thics of* I*ntensification* (pp. 191–198). International Library of Environmental, Agricultural and Food Ethics, vol. 16. Springer. https://doi.org/10.1007/978-1-4020-8722-6_13

Sandøe, P., Palmer, C., Corr, S., Astrup, A., & Bjørnvad, C.R. (2014). Canine and feline obesity: A One Health perspective. *Veterinary Record, 175*(24), 610–616.

Savage-Rumbaugh, S., McDonald, K., Sevcik, R.A., Hopkins, W.D., & Rubert, E. (1986). Spontaneous symbol acquisition and communicative use by pygmy chimpanzees (*Pan paniscus*). *Journal of Experimental Psychology: General, 115*(3), 211.

Scientific Committee on Animal Health and Welfare (SCAHAW). (1999). https://food.ec.europa.eu/horizontal-topics/expert-groups/scientific-committees/scientific-committee-animal-health-and-animal-welfare-archive_en

Schaffner, J. (2011). *An Introduction to Animals and the Law*. Palgrave Macmillan.

Schalke, E., Stichnoth, J., Ott, S., & Jones-Baade, R. (2007). Clinical signs caused by the use of electric training collars on dogs in everyday life situations. *Applied Animal Behaviour Science, 105*(4), 369–380. https://doi.org/10.1016/j.applanim.2006.11.002

Schmied, C., Boivin, X., Scala, S., & Waiblinger, S. (2010). Effect of previous stroking on reactions to a veterinary procedure: Behaviour and heart rate of dairy cows. *Interaction Studies, 11*(3), 467–481. https://doi.org/10.1075/is.11.3.08sch

Schroepfer, K.K., Rosati, A.G., Chartrand, T., & Hare, B. (2011). Use of 'entertainment' chimpanzees in commercials distorts public perception regarding their conservation status. *PLOS ONE, 6*(1), e26048.

Schuck, S.E., Emmerson, N.A., Fine, A.H., & Lakes, K.D. (2015). Canine-assisted therapy for children with ADHD: Preliminary findings from the positive assertive cooperative kids study. *Journal of Attention Disorders, 19*(2), 125–137. https://doi.org/10.1177/1087054713502080

Seabrook, M.F. (1972a). A study to determine the influence of the herdsmen's personality on milk yield. *Journal of Agriculture Labour Science, 1*, 45–59.

Seabrook, M.F. (1972b). *A Study on the Influence of the Cowman's Personality and Job Satisfaction on Milk Yield of Dairy Cows*. Joint Conference of the British Society for Agriculture Labour Science and the Ergonomics Research Society, National College of Agricultural Engineering, UK, September, 1972.

Seabrook, M.F. (1984). The psychological interaction between the stockman and his animals and its influence on performance of pigs and dairy cows. *Veterinary Record, 115*(4), 84.

Seabrook, M.F. (1996). *The Role of the Human Factor in Increasing Performance and Profitability in Pig Production*. Filozzo Rhone Poulenc Conference, Moderna, Italy, pp. 2–13.

Sekhar, N.U. (2003). Local people's attitudes towards conservation and wildlife tourism around Sariska Tiger Reserve, India. *Journal of Environmental Management, 69*(4), 339–347.

Sendler, D.J. (2018). Why people who have sex with animals believe that it is their sexual orientation – a grounded theory study of online communities of zoophiles. *Deviant Behavior, 39*(11), 1507–1514.

Serpell, J. (2003). Anthropomorphism and anthropomorphic selection – beyond the 'cute response'. *Society & Animals, 11*(1), 83–100.

Serpell, J. (2011). *Human–Dog Relationships Worldwide*. Proceedings of the Expert Meeting on Dog Population Management, Banna, Italy, 15–19 March, 2011.

Serpell, J.A. (2004). Factors influencing human attitudes to animals and their welfare. *Animal Welfare, 13*, 145–151.

Serpell, J.A. (2010). Animal-assisted interventions in historical perspective. In A.H. Fine (Ed.), *Animal-Assisted Therapy: Foundations and Guidelines for Animal-Assisted Interventions* (pp. 17–32). Elsevier.

Serpell, J.A., Kruger, K.A., Freeman, L.M., Griffin J.A., & Ng, Z.Y. (2020). Current standards and practices within the therapy dog industry: Results of a representative survey of United States therapy dog organizations. *Frontiers in Veterinary Science, 7*, 35.

Seto, M.C., Curry, S., Dawson, S.J., Bradford, J.M., & Chivers, M.L. (2021). Concordance of paraphilic interests and behaviors. *Journal of Sex Research, 58*(4), 424–437. https://doi.org/10.1080/00224499.2020.1830018

Seto, M.C., & Lalumiere, M.L. (2010). What is so special about male adolescent sexual offending? A review and test of explanations through meta-analysis. *Psychological Bulletin, 136*(4), 526–575. https://doi.org/10.1037/a0019700

Seyfarth, R.M., & Cheney, D.L. (1993). Meaning, reference, and intentionality in the natural vocalizations of monkeys. In H.L. Roitblat, L.M. Herman, P.E. Nachtigall (Eds.), *Language and Communication: Comparative Perspectives* (pp. 195–219). Lawrence Erlbaum.

Seyfarth, R.M., Cheney, D.L., & Marler, P. (1980). Vervet monkey alarm calls: Semantic communication in a free-ranging primate. *Animal Behaviour, 28*(4), 1070–1094.

Shah, S.Z.A., Nawaz, Z., Nawaz, S., Carder, G., Ali, M., Soomro, N., & Compston, P.C. (2019). The role and welfare of cart donkeys used in waste management in Karachi, Pakistan. *Animals, 9*(4), 159.

Shea, B.T. (1989). Heterochrony in human evolution: The case for neoteny reconsidered. *American Journal of Physical Anthropology, 32*(S10), 69–101.

Shipman, P. (Ed.). (2010). The animal connection and human evolution. *Current Anthropology, 51*(4), 519–538.

Shipman, P. (2011). *The Animal Connection: A New Perspective on What Makes us Human*. W.W. Norton.

Shipman, P. (2015). *The Invaders*. Harvard University Press.

Shipman, P. (2021). *Our Oldest Companions: The Story of the First Dogs*. Harvard University Press.

Silva, K., Correia, R., Lima, M., Magalhães, A., & de Sousa, L. (2011). Can dogs prime autistic children for therapy? Evidence from a single case study. *Journal of Alternative and Complementary Medicine, 17*(7), 1–5. https://doi.org/10.1089/acm.2010.0436

Simion, F., Regolin, L., & Buff, H. (2008). A pre-disposition for biological motion in the newborn baby. *Proceedings of the National Academy of Sciences, 105*, 809–813.

Simmons, C.A., & Lehmann, P. (2007). Exploring the link between pet abuse and controlling behaviors in violent relationships. *Journal of Interpersonal Violence, 22*, 1211–1222. https://doi.org/10.1177/0886260507303734

Singh, P., & Kaur, M. (2015). Sniffers as an aid in crime investigation. *New York Science Journal, 8*(10), 87–92.

Siniscalchi, M., D'Ingeo, S., Minunno, M., & Quaranta, A. (2018). Communication in dogs. *Animals, 8*(8), 131. https://doi.org/10.3390/ani8080131

Siniscalchi, M., d'Ingeo, S., & Quaranta, A. (2021). Lateralized emotional functioning in domestic animals. *Applied Animal Behaviour Science*, 105282.

Situ, Y., & Emmons, D. (2000). *Environmental Crime*. SAGE.

Skippen, L., Collier, J., & Kithuka, J.M. (2021).The donkey skin trade: A growing global problem. *Brazilian Journal of Veterinary Research and Animal Science , 58*, e175262. https://doi.org/10.11606/issn.1678-4456.bjvras.2021.175262.

Skogen, K. (2017). Unintended consequences in conservation: How conflict mitigation may raise the conflict level – the case of wolf management in Norway. In C.M. Hill, A.D. Webber, & N.E.C. Priston (Eds.), *Understanding conflicts About Wildlife: A biosocial Approach* (pp. 49–64). Berghahn Books.

Smart, J.J.C., & Williams, B. (1973). *Utilitarianism: For and Against*. Cambridge University Press.

Smith, D.W., & Peterson, R.O. (2021). Intended and unintended consequences of wolf restoration to yellowstone and isle royale national parks. *Conservation Science and Practice, 3*(4), e413.

Smith, J.J. (2014). Human–animal relationships in zoo-housed orangutans (*p. abelii*) and gorillas (*g. g. gorilla*): The effects of familiarity: Familiarity, hars, and ape-human interaction. *American Journal of Primatology, 76*(10), 942–955. https://doi.org/10.1002/ajp.22280

Smith, P., & Daniel, C. (2000). *The Chicken Book*. University of Georgia Press.

Smith, S.M., & Vale, W.W. (2006). The role of the hypothalamic-pituitary-adrenal axis in neuroendocrine responses to stress. *Dialogues in Clinical Neuroscience, 8*(4), 383–395. www.dialogues-cns.org

Smolkovic, I., Fajfar, M., & Mlinaric, V. (2012). Attachment to pets and interpersonal relationships: Can a four-legged friend replace a two-legged one? *Journal of European Psychology Students, 3*(1), 15–23. https://doi.org/10.5334/jeps.ao

Sneddon, L.U. (2019). Evolution of nociception and pain: Evidence from fish models. *Philosophical Transactions of the Royal Society B, 374*(1785), 20190290.

Sneddon, L.U., Elwood, R.W., Adamo, S.A., & Leach, M.C. (2014). Defining and assessing animal pain. *Animal Behaviour, 97*, 201–212.

Sollund, R. (2013). Animal trafficking and trade: Abuse and species injustice. In R. Walters, D. Westerhuis, & T. Wyatt (Eds.), *Emerging Issues in Green Criminology:* Exploring power, justice and harm (pp. 72–92). Palgrave Macmillan.

Somervill, J.W., Swanson, A.M., Robertson, R.L., Arnett, M.A., & MacLin, O.H. (2009). Handling a dog by children with attention- deficit/hyperactivity disorder: Calming or exciting? *North American Journal of Psychology, 11*(1), 111–120.

Somppi, S., Törnqvist, H., Hänninen, L., Krause, C.M., & Vainio, O. (2014). How dogs scan familiar and inverted faces: An eye movement study. *Animal Cognition, 17*, 793–803. https://doi.org/10.1007/s10071-013-0713-0

Sousa, J., Hill, C.M., & Ainslie, A. (2017). Chimpanzees, sorcery and contestation in a protected area in Guinea Bissau. *Social Anthropology, 25*, 364–379.

South, N., & Wyatt, T. (2011). Comparing illicit trades in wildlife and drugs: An exploratory study. *Deviant Behavior, 32*(6), 538–561.

Spitznagel, M.B., Jacobson, D.M., Cox, M.D., & Carlson, M.D. (2017). Caregiver burden in owners of a sick companion animal: A cross-sectional observational study. *Veterinary Record, 181*(12), 321. https://doi.org/10.1136/vr.104295

St. John, F.A.V., Steadman, J., Austen, G., & Redpath, S.M. (2019). Value diversity and conservation conflict: Lessons from the management of red grouse and hen harriers in England. *People and Nature, 1*, 6–17.

Stephens, T. (Ed.). (2021). *One Welfare in Practice: The Role of the Veterinarian*. CRC Press.

Stern, A.W., & Smith-Blackmore, M. (2016). Veterinary forensic pathology of animal sexual abuse. *Veterinary Pathology, 53*(5), 1057–1066. https://doi.org/10.1177/0300985816643574

Stiffelman, B. (2019). No longer the gold standard: Probabilistic genotyping is changing the nature of DNA evidence in criminal trials. *Berkeley Journal of Criminal Law, 24*, 110.

Strand, E.B., & Faver, C.A. (2006). Battered women's concern for their pets: A closer look. *Journal of Family Social Work, 9*, 39–58. https://doi.org/10.1300/J039v09n04_04

Stringer, A., Lunn, D.P., & Reid, S. (2015). Science in brief: Report on the first Havemeyer workshop on infectious diseases in working equids, Addis Ababa, Ethiopia, November 2013. *Equine Veterinary Journal, 47*(1), 6–9. https://doi.org/10.1111/evj.12359

Strong, M., & Silva, J.A. (2020). Impacts of hunting prohibitions on multidimensional well-being. *Biological Conservation, 243*, 108451. https://doi.org/10.1016/j.biocon.2020.108451

Su, B., & Martens, P. (2017). Public attitudes toward animals and the influential factors in contemporary China. *Animal Welfare, 26*(2), 239–247. https://doi.org/ 10.7120/09627286.26.2.239

Sueur, C., Zanaz, S., & Pelé, M. (2022). Incorporating animal agency into research design could improve behavioral and neuroscience research. *Journal of Comparative Psychology, 137*(2), 129.

Sumner, R.C., & Goodenough, A.E. (2020). A walk on the wild side: How interactions with non-companion animals might help reduce human stress. *People and Nature, 2*(2), 395–405.

Sundman, A.S., Poucke, E. Van, Holm, A.C.S., Faresjö, Å., Theodorsson, E., Jensen, P. et al. (2019). Long-term stress levels are synchronized in dogs and their owners. *Scientific Reports, 9*(1), 1–7. https://doi.org/10.1038/s41598-019-43851-x

Sutherland, E. (1973). *On analysing Crime*. University of Chicago Press

Suzman, J. (2014, 31 August). Sympathy for a desert dog. *New York Times*. Opinion Pages. https://opinionator.blogs.nytimes.com/2014/08/31/sympathy-for-a-desert-dog/

Suzuki, T.N., Griesser, M., & Wheatcroft, D. (2019). Syntactic rules in avian vocal sequences as a window into the evolution of compositionality. *Animal Behaviour, 151*, 267–274.

Suzuki, T.N., Wheatcroft, D., & Griesser, M. (2016). Experimental evidence for compositional syntax in bird calls. *Nature Communications, 7*(1), 10986.

Suzuki, T.N., Wheatcroft, D., & Griesser, M. (2018). Call combinations in birds and the evolution of compositional syntax. *PLOS Biology, 16*(8), e2006532.

Sykes, G.M., & Matza, D. (1957). Techniques of neutralization: A theory of delinquency. *American Sociological Review, 22*, 664–673.

Sykes, N. (2012). A social perspective on the introduction of exotic animals: The case of the chicken. *World Archaeology, 44*(1), 158–169.

Szymańska-Pytlińska, M.E., Beisert, M.J., & Słopień, A.U. (2021). Development of zoophilic interests and behaviors in the example of an adolescent male. *Current Issues in Personality Psychology, 9*(1), 26–36. https://czasopisma.bg.ug.edu.pl/index.php/CIiPP/article/view/5762

Tam, K.P. (2013). Concepts and measures related to connection to nature: Similarities and differences. *Journal of Environmental Psychology, 34*, 64–78. https://doi.org/10.1016/j.jenvp.2013.01.004

Tarazona, A.M., Ceballos, M.C., & Broom, D.M. (2020). Human relationships with domestic and other animals: One health, one welfare, one biology. *Animals, 10*(1), 43.

Terrace, H.S., Petitto, L.A., Sanders, R.J., & Bever, T.G. (1979). Can an ape create a sentence? *Science, 206*(4421), 891–902.

Thayer, E.R., & Stevens, J.R. (2022). Effects of human-animal interactions on affect and cognition. *Human-Animal Interaction Bulletin, 10*(2), 73–98.

Thielke, L.E., & Udell, M.A.R. (2017). The role of oxytocin in relationships between dogs and humans and potential applications for the treatment of separation anxiety in dogs. *Biological Reviews, 92*(1), 378–388. https://doi.org/10.1111/brv.12235

Tiira, K., & Lohi, H. (2015). Early life experiences and exercise associate with canine anxieties. *PLOS ONE, 10*(11), e0141907.

Tonoike, A., Terauchi, G., Inoue-Murayama, M., Nagasawa, M., Mogi, K., & Kikusui, T. (2016). The frequency variations of the oxytocin receptor gene polymorphisms among dog breeds. Journal of Azabu University, 27, 11–18.

Touroo, R., & Fitch, A. (2016). Identification, collection, and preservation of veterinary forensic evidence: On scene and during the postmortem examination. *Veterinary Pathology, 53*(5), 880-887.

Townsend, S.W., Engesser, S., Stoll, S., Zuberbühler, K., & Bickel, B. (2018). Compositionality in animals and humans. *PLOS Biology, 16*(8), e2006425.

Treves, A., Naughton-Treves, L., Harper, E.K., Mladenoff, D.J., Rose, R.A., Sickley, T.A. et al. (2004). Predicting human-carnivore conflict: A spatial model derived from 25 years of data on wolf predation on livestock. *Conservation Biology, 18*(1), 114–125.

Trotter, K.S., Chandler, C.K., Goodwin-Bond, D., & Casey, J. (2008). A comparative study of the efficacy of group equine assisted counselling with at-risk children and adolescents. *Journal of Creativity in Mental Health, 3*, 254–284. https://doi.org/10.1080/15401380802356880

Trut, L., Oskina, I., & Kharlamova, A. (2009). Animal evolution during domestication: The domesticated fox as a model. *BioEssays, 31*, 349–360.

Trut, L.N., Kharlamova, A.V., Pilipenko, A.S., Herbeck, Y.E. (2021).The fox domestication experiment and dog evolution: A view based on modern molecular, genetic and archaeological data. *Russian Journal of Genetics, 57*, 778–794.

Turcsán, B., Range, F., Virányi, Z., Miklósi, Á., & Kubinyi, E. (2012). Birds of a feather flock together? Perceived personality matching in owner–dog dyads. *Applied Animal Behaviour Science, 140*(3–4), 154–160. https://doi.org/10.1016/J.APPLANIM.2012.06.004

Ujita, A., Faro, L.E., Vicentini, R.R., Lima, M.P., Fernandes, L.O., Oliveira, A.P., Veroneze, R., & Negrao, J. (2020). Effect of positive tactile stimulation and prepartum milking routine training on behaviour, cortisol and oxytocin in milking, milk composition, and milk yield in Gyr cows in early lactation. *Applied Animal Behaviour Science, 234*, 105205

UK Government. (n.d., a). Animals to be formally recognised as sentient beings in domestic law. www.gov.uk

UK Government. (n.d., b). List of zoonotic diseases. www.gov.uk/government/publications/list-of-zoonotic-diseases/list-of-zoonotic-diseases

United Nations. (2020). Sustainable development goals. www.un.org/sustainabledevelopment/development-agenda/

United Nations Office on Drugs and Crime (UNODC). (2004). United Nations convention against transnational organized crime and the protocols thereto. www.unodc.org/documents/treaties/UNTOC/Publications/TOC%20Convention/TOCebook-e.pdf

United Nations World Tourism Organization (UNWTO). (2015). Towards measuring the economic value of wildlife watching tourism in Africa. https://sustainabledevelopment.un.org/content/documents/1882unwtowildlifepaper.pdf

Urquiza-Haas, E.G., & Kotrschal, K. (2015). The mind behind anthropomorphic thinking: Attribution of mental states to other species. *Animal Behaviour, 109*, 167–176.

Usui, R., & Funck, C. (2017). Not quite wild, but not domesticated either: Contradicting management decisions on free-ranging sika deer (*Cervus nippon*) at two tourism sites in Japan. In I. Borges de Lima & R.J. Green (Eds.), *Wildlife Tourism, Environmental Learning and Ethical Encounters* (pp. 247–261). Geoheritage, Geoparks and Geotourism. Springer.

Usui, R., Sheeran, L.K., Li, J., Sun, L., Wang, X., Pritchard, A.J., DuVall-Lash, A.S., & Wagner, R.S. (2014). Park rangers' behaviors and their effects on tourists and Tibetan macaques (Macaca thibetana) at Mt. Huangshan, China. *Animals, 4*, 546–561.

Valette, D. (2014, May). *Invisible Helpers. Women's Views on the Contributions of Working Donkeys, Horses, and Mules to* Their Lives. Brooke. www.thebrooke.org/sites/default/files/Advocacy-and-policy/Invisible-helpers-voices-from-women.pdf

Valstad, M., Alvares, G.A., Egknud, M., Matziorinis, A.M., Andreassen, O.A., Westlye, L.T. et al. (2017). The correlation between central and peripheral oxytocin concentrations: A systematic review and meta-analysis. *Neuroscience & Biobehavioral Reviews, 78*, 117–124.

van der Zee, E., & Weary, D. (2010). Communication. In D.S. Mills & J.N. Marchant-Forde (Eds.), *The Encyclopedia of Applied Animal Behaviour and Welfare*. CABI.

van der Zee, E., Zulch, H., & Mills, D. (2012). Word generalization by a dog (*Canis familiaris*): Is shape important? *PLOS ONE*, 7(11), e49382. https://doi.org/10.1371/journal. pone.0049382

van Wijk, A., Hardeman, M., & Endenburg, N. (2018). Animal abuse: Offender and offence characteristics. A descriptive study. *Journal of Investigative Psychology and Offender Profiling, 15*(2), 175–186. https://doi.org/10.1002/jip.1499

Varner, G. (2011). Environmental ethics, hunting and the place of animals. In T.L. Beauchamp & R.G. Frey (Eds.), *The Oxford handbook of Animal Ethics* (pp. 854–876). Oxford University Press.

Vaughn, M.G., Fu, Q., DeLisi, M., Beaver, K.M., Perron, B.E., Terrell, K. et al. (2009). Correlates of cruelty to animals in the United States: Results from the national epidemiologic survey on alcohol and related conditions. *Journal of Psychiatric Research, 43*, 1213–1218. https://doi.org/10.1016/j.jpsychires.2009.04.011

Ventura C., Denton E., & Van Court E. (2021). Principles of death and dying. In *The Emergency Medical Responder*: Training and Succeeding as an EMT/EMR (pp. 119–120). Springer, Cham. https://doi.org/10.1007/978-3-030-64396-6_12

Viau, R., Arsenault-Lapierre, G., Fecteau, S., Champagne, N., Walker, C.D., & Lupien, S. (2010). Effect of service dogs on salivary cortisol secretion in autistic children. *Psychoneuroendrocrinology, 35*(8), 1187–1193. https://doi.org/10.1016/j.psyneuen.2010.02.004

Vigne, J.D. (2011). The origins of animal domestication and husbandry: A major change in the history of humanity and the biosphere. *Comptes Rendus Biologies, 334*, 171–181.

Vogel, A., Mello, M.A.D.S., de Barros, J.F.P., & Lody, R. (1993). *Galinha d'Angola: Iniciação e Identidade na Cultura Afro-Brasileira* (3rd ed.). Pallas.

Volk, J.O., Schimmack, U., Strand, E.B., Reinhard, A., Vasconcelos, J., Hahn, J. et al. (2022). Executive summary of the merck animal health veterinarian well-being study III and veterinary support staff study. *Journal of the American Veterinary Medical Association, 260*(12), 1547–1553. https://doi.org/10.2460/javma.22.03.0134

Volsche, S. (2018). Negotiated bonds: The practice of childfree pet parenting. *Anthrozoös, 31*(3), 367–377. https://doi.org/10.1080/08927936.2018.1455470

Volsche, S. (2021). Pet parenting in the United States: Investigating an evolutionary puzzle. *Evolutionary Psychology, 19*(3), 1–9. https://doi.org/10.1177/14747049211038297

Volsche, S., Mohan, M., Gray, P.B., & Rangaswamy, M. (2019). An exploration of attitudes toward dogs among college students in Bangalore, India. *Animals, 9*, 514. https://doi.org/10.3390/ani9080514

Volsche, S., Mukherjee, R., & Rangaswamy, M. (2021). The difference is in the details: Attachment and cross-species parenting in the United States and India. *Anthrozoös, 35*(3), 398–408. https://doi.org/10.1080/08927936.2021.1996026

von Essen, E., & Allen, M.P. (2017). Reconsidering illegal hunting as a crime of dissent: Implication for justice and deliberative uptake. *Criminal Law, Philosophy, 11*, 213–228. https://doi.org/10.1007/s11572-014-9364-8

von Holdt, B.M., Shuldiner, E., Koch, I., Kartzinel, R., Hogan, A., Brubaker, L. et al. (2017). Structural variants in genes associated with human Williams-Beuren syndrome underlie stereotypical hypersociability in domestic dogs. *Science Advances, 3*(7), e1700398.

Voznesenskiy, S., Rivera-Quinatoa, J.A., Bonilla-Yacelga, K.A., & Cedeño-Zamora, M.N. (2016). Do equine-assisted physical activities help to develop gross motor skills in children with the down syndrome? Short-term results. *Procedia Social and Behavioural Science, 233*, 307–312. https://doi.org/10.1016/j.sbspro.2016.10.140

Waddington, C.H. (1953). Genetic assimilation of an acquired character. *Evolution, 7*, 118–126.

Waiblinger, S., Boivin, X., Pedersen, V., Tosi, M.V., Janczak, A.M., Visser, E.K. et al. (2006). Assessing the human–animal relationship in farmed species: A critical review. *Applied Animal Behaviour Science, 101*, 185–242.

Waiblinger, S., Menke, C., & Coleman, G.J. (2002). The relationship between attitudes, personal characteristics and behaviour of stockpeople and subsequent behaviour and production of dairy cows. *Applied Animal Behaviour Science, 79*, 195–219.

Waiblinger, S., Menke, C., Korff, J., & Bucher, A. (2004). Previous handling and gentle interactions affect behaviour and heart rate of dairy cows during a veterinary procedure. *Applied Animal Behaviour Science, 85*, 31–42.

Walsh, F. (2009). Human–animal bonds I: The relational significance of companion animals. *Family Process, 48*(4), 462–480.

Walters, G.D. (2013). Testing the specificity postulate of the violence graduation hypothesis: Meta-analyses of the animal cruelty-offending relationship. *Aggression and Violent Behavior, 18*, 797–802. https://doi.org/10.1016/j.avb.2013.10.002

Ward, S.C., Whalon, K., Rusnak, K., Wendell, K., & Paschall, N. (2013). The association between therapeutic horseback riding and the social communication and sensory reactions of children with Autism. *Journal of Autism and Developmental Disorders, 43*, 2190–2198.

Ward, S.J., & Melfi, V. (2015). Keeper–animal interactions: Differences between the behaviour of zoo animals affect stockmanship. *PLOS ONE, 10*(10), e0140237. https://doi.org/10.1371/journal.pone.0140237

Ward-Griffin, E., Klaiber, P., Collins, H.K., Owens, R.L., Coren, S., & Chen, F.S. (2018). Petting away pre-exam stress: The effect of therapy dog sessions on student well-being. *Stress and Health, 34*(5). https://doi.org/10.1002/smi.2804

Warwick, C., Steedman, C., Jessop, M., Arena, P., Pilny, A., & Nicholas, E. (2018). Exotic pet suitability: Understanding some problems and using a labeling system to aid animal welfare, environment, and consumer protection. *Journal of Veterinary Behavior, 26*, 17–26.

Warwick, C., Steedman, C., Jessop, M., Toland, E., & Lindley, S. (2014). Assigning degrees of ease or difficulty for pet animal maintenance: The EMODE system concept. *Journal of Agricultural and Environmental Ethics, 27*(1), 87–101.

Waters, S., Setchell, J.M., Maréchal, L., Oram, F., & Cheyne, S.M. (2021). *Best Practice Guidelines for Responsible Images of Non-Human Primates*. International Union for the Conservation of Nature. Switzerland Press.

Waytz, A., Cacioppo, J., & Epley, N. (2010). Who sees human? The stability and importance of individual differences in anthropomorphism. *Perspectives on Psychological Science, 5*(3), 219–232.

Webb, A.A., & Cullen, C.L. (2010). Coat color and coat color pattern-related neurologic and neuro-ophthalmic diseases. *Canadian Veterinary Journal, 51*(6), 653.

Weir, B., Crisp, D., O'Dell, C.W., Basu, S., Chatterjee, A., Kolassa, J. et al. (2021). Regional impacts of COVID-19 on carbon dioxide detected worldwide from space. *Science Advances, 7*(45), eabf9415.

Weldegebriel, Z.B., & Amphune, B.E. (2017). Livelihood resilience in the face of recurring floods: An empirical evidence from Northwest Ethiopia. *Geoenvironmental Disasters, 4*(1), 1–19.

Wells, D.L. (2009). The effects of animals on human health and well-being. *Journal of Social Issues, 65*, 523–543.

Wheeler, B.C. (2009). Monkeys crying wolf? Tufted capuchin monkeys use anti-predator calls to usurp resources from conspecifics. *Proceedings of the Royal Society B: Biological Sciences, 276*(1669), 3013–3018.

White, R., & Heckenberg, D. (2014). *Green Criminology: An Introduction to the Study of Environmental Harm*. Routledge

Wielebnowski, N.C. (1999). Behavioral differences as predictors of breeding status in captive cheetahs. *Zoo Biology, 18*(4), 335–349.

Wilkins, A.S., Wrangham, R., & Fitch, W.T. (2021). The neural crest/syndrome hypothesis, explained: Reply to M. Johnsson, R. Henrioksen, and D. Wright. *Genetics, 219*(1), iyab098.

Wilkins, A.S., Wrangham, R.W., & Fitch, W.T. (2014). The 'Domestication syndrome' in mammals: A unified explanation based on neural crest cell behavior and genetics. *Genetics, 197*, 795–808.

Williams, C.J., & Weinberg, M.S. (2003). Zoophilia in men: A study of sexual interest in animals. *Archives of Sexual Behavior, 32*(6), 523–535. https://doi.org/10.1023/A:1026085410617

Williamson, E.A., & Feistner, A.T.C. (2011). Habituating primates: Processes, techniques, variables and ethics. In J.M. Setchell & D.J. Curtis (Eds.), *Field and Laboratory Methods in Primatology: A Practical Guide* (2nd ed., pp. 33–50). https://doi.org/10.1017/CBO9780511921643.004

Wilson, E.O. (1984). *Biophilia*. Harvard University Press.

Wirobski, G., Range, F., Schaebs, F.S., Palme, R., Deschner, T., & Pescini, S.M. (2021). Life experience rather than domestication accounts for dogs' increased oxytocin release during social contact with humans. *Scientific Reports, 11*(14423), 1–12. https://doi.org/10.1038/s41598-021-93922-1

Working Together to Transform the Brick Kiln Industry. Brooke. www.thebrooke.org/sites/default/files/Professionals/Brooke_Brick_Collaborative_Brief.pdf

World Health Organization (WHO). www.who.int/news-room/fact-sheets/detail/animal-bites

World Tourism Organization/Guangdong Chimelong Group. (2020). *Sustainable Development of Wildlife Tourism in Asia and the Pacific*. https://doi.org/10.18111/9789284421572

Wright, D., Henriksen, R., & Johnsson, M. (2020). Defining the domestication syndrome: Comment on Lord et al, 2020. *TREE, 35*, 1059–1060.

Wright, J.C. (1996). Canine aggression: Dog bites to people. In V. Voith & P.L. Borchelt (Eds.), *Readings in Companion Animal Behavior* (pp. 240–246). Veterinary Learning Systems.

Wyatt, T. (2013). *Wildlife Trafficking: A Deconstruction of the Crime, the Victims and the Offenders*. Palgrave Macmillan.

Wyatt, T., van Uhm, D., & Nurse, A. (2020). Differentiating criminal networks in the illegal wildlife trade: Organized, corporate and disorganized crime. *Trends in Organised Crime, 33*, 350–366. https://doi.org/10.1007/s12117-020-09385-9

Xiang, L., Zhou, G., Su, P., Xia, S., Han, B., Wang, Y. et al. (2013). Could postmortem hemorrhage occur in the brain? A preliminary study on the establishment and investigation of postmortem hypostatic hemorrhage using rabbit models. *American Journal of Forensic Medicine and Pathology, 34*(2), 147–149.

Yamada, T. (2018). The Ainu bear ceremony and the logic behind hunting the deified bear. *Journal of Northern Studies, 12*(1), 35–51.

Yatcilla, J.K. (2021). A panorama of human–animal interactions research: Bibliometric analysis of HAI articles 1982–2018. *Anthrozoös, 34*, 161–173.

Yoshimoto, K., Ueda, S., Kitamura, Y., Inden, M., Hattori, H., Ishikawa, N. et al. (2012). Administration of rotenone enhanced voluntary alcohol drinking behavior in C57BL/6J mice. *Legal Medicine, 14*(5), 229–238.

Young, R. (2017). *The Secret Life of Cows*. Faber & Faber.

Zebrowitz, L.A., Brownlow, S., & Olson, K. (1992). Baby talk to the babyfaced. *Journal of Nonverbal Behavior, 16*(3), 143–158.

Zeder, M.A. (2015). Core questions in domestication research. *Proceedings of the National Academy of Sciences of the United States of America, 112*, 3191–3198.

Zheng, X., Meng, X., & Ji, Y. (2018). Intentional inference during infants' observational word learning. *Lingua, 207*, 38–48.

Zilcha-Mano, S., Mikulincer, M., & Shaver, P.R. (2011). An attachment perspective on human–pet relationships: Conceptualization and assessment of pet attachment orientations. *Journal of Research in Personality, 45*(4), 345–357.

Zilcha-Mano, S., Mikulincer, M., & Shaver, P.R. (2012). Pets as safe havens and secure bases: The moderating role of pet attachment orientations. *Journal of Research in Personality, 46*(5), 571–580.

Zimen, E. (1997). *Der Wolf* [The wolf]. Kosmos.

Zimmermann, H., Blažek, R., Polačik, M., & Reichard, M. (2022). Individual experience as a key to success for the cuckoo catfish brood parasitism. *Nature Communications, 13*(1), 1–9.

Zisook, S., & Shear, K. (2009). Grief and bereavement: What psychiatrists need to know. *World Psychiatry, 8*(2), 74. https://doi.org/10.1002/J.2051-5545.2009.TB00217.X

Ziv, G. (2017). The effects of using aversive training methods in dogs – a review. *Journal of Veterinary Behavior, 19*, 50–60.

Zoubek, E. (2017). *From Egg to Dead. Small Scale Chicken Keeping in Great Britain*. [Unpublished doctoral thesis]. University of Roehampton (London).

Zubedat, S., Aga-Mizrachi, S., Cymerblit-Sabba, A., Shwartz, J., Leon, J.F., Rozen, S. et al. (2014). Human–animal interface: The effects of handler's stress on the performance of canines in an explosive detection task. *Applied Animal Behaviour Science, 158*, 69–75.

Zuberbühler, K. (2019). Evolutionary roads to syntax. *Animal Behaviour, 151*, 259–265.
Zuberbühler, K. (2020). Syntax and compositionality in animal communication. *Philosophical Transactions of the Royal Society B, 375*(1789), 20190062.
Zuberbühler, K. (2021). Event parsing and the origins of grammar. *Wiley Interdisciplinary Reviews: Cognitive Science*, e1587.
Żukiewicz-Sobczak, W.A., Chmielewska-Badora, J., Wróblewska, P., & Zwoliński, J. (2013). Farmers' occupational diseases of allergenic and zoonotic origin. *Advances in Dermatology and Allergology/ Postępy Dermatologii i Alergologii, 30*(5), 311–315.

Index

ANIMAL INDEX

GENERAL INDEX

For Product Safety Concerns and Information please contact our EU
representative GPSR@taylorandfrancis.com
Taylor & Francis Verlag GmbH, Kaufingerstraße 24, 80331 München, Germany

www.ingramcontent.com/pod-product-compliance
Ingram Content Group UK Ltd.
Pitfield, Milton Keynes, MK11 3LW, UK
UKHW030828080625
459435UK00014B/590